Testi
W!
afte

C0001138722

"How wonderful that Tom has chosen to share ㄴ... ... ᵤ and life experiences. We met Tom seventeen years ago during our worst of times and we treasure his wisdom and friendship to this day."

~MICHAEL AND DEBORAH HUTCHINSON
Parents of SPC Ray Joseph Hutchinson,
KIA Mosul Iraq 12/7/03

"As a consummate and well-decorated leader, General Bostick possesses not only firsthand knowledge about leading others with resilience through accomplishments and setbacks, he also has an ability to share meaningful information in a way that it can be heard and consumed. He was a keynote speaker at our National Practice Conference for psychologists and engaged us with his warmth, humor, insight, and expertise. Readers of *Winning After Losing, Strategies for Successful Leadership* will no doubt feel renewed and supported, not to mention equipped with strategies for becoming resilient leaders themselves."

~JANA N. MARTIN,
PhD, CEO, The Trust

"My time in Bravo Company, 54th Engineer Battalion, with then Captain Bostick, was the most influential period in my twenty-eight-year Army career and more importantly, my life; he took a young kid with no direction and gave me purpose, leadership, and an unshakable foundation that supports me to this day. My assignment to Bravo Company in the early '80s was both exciting and rewarding, as we were only eleven kilometers from the Fulda Gap during the Cold War. Bravo Company was the best of two worlds; we worked hard and played hard. We conducted numerous training exercises to hone our engineering skills and won numerous awards and competitions, most notably, the Army's Best Maintenance Award; however, we enjoyed ourselves too, with incredible company trips to Italy, Spain and throughout Germany . . . it was a great time!"

~SERGEANT MAJOR SCOTT KUHAR,
U.S. Army

"I took these same B Company team building maintenance programs and skills into company command at 101st. Using the same vehicle inspection checklists created in the 54th, each officer was trained on how to use them. They in turn trained their drivers and NCOs. It really paid off in deployment readiness since we had to load planes or rig vehicles for helicopters on short notice."

~STEVE DUNHAM,
served as Executive Officer in B Company,
54th Engineer Battalion

"Tom Bostick is one of the finest people I have ever met. Whether competing in triathlons, speaking to fourth graders, or directing the cleanup of Super Storm Sandy, his leadership in every aspect is envied by those who have been blessed to meet him. Always a gentleman and a great Commander, he communicates at all levels so there is no possible way to misunderstand. The Army is a better institution because of LTG (Ret) Tom Bostick's thirty-eight years of service. It was a privilege to have worked for him."

~KAREN HUFF,
Executive Assistant to Lt. Gen. Bostick for over ten years

"LTG Bostick is an inclusive leader who leveraged the power of the entire team to accomplish extraordinary feats. He provided me the opportunity and, more importantly, the belief that my contributions were valuable. This has had a profound effect upon my leadership style, ensuring everyone, regardless of their role or position, is a valued contributor to my team."

~COLONEL ROSS A. DAVIDSON, U.S. ARMY.
The original architect of the Davidson-Style Sea Hut

"Working with then Colonel Thomas Bostick to build Tuzla, Bosnia and Herzegovina airfield into a strategic platform was memorable. He is remembered as a unique leader with multifaceted attributes that makes him one-off. At first, I did not know what to expect of Colonel Bostick because all I knew was that I had to accompany him to Tuzla to build a runway and an aircraft parking apron to support large-frame aircraft in what I thought was an impossible time. I quickly noted he was more than just the head engineer; he was a leader who is results oriented, humble, and visible. He took care of people as they worked to accomplish their mission, leading with emotional intelligence and intellect. I remember saying in 1998 to someone in Tuzla, 'He is a general officer'!"

~AIR FORCE MASTER SERGEANT PATRICK DAIZE,
Project Manager, Tuzla Strategic Airfield Construction

"General Tom Bostick gave his leaders the finest gift possible from another leader—empowerment. It was the trust and freedom to experiment that allowed for the team recruiting concept to work. Soldiers from all walks of life gel into teams and naturally migrate to those things they do best. General Bostick inherently knew soldiers thrive in a team environment where their individual talents together bring out the finest in themselves and allow the mission to get accomplished. It only made sense that Team Recruiting would be successful using a structure our soldiers knew innately and understood. The effort would have never been possible without the personal leadership of General Bostick. It was a pleasure serving in his command team."

~LT. COL. (RET.) TED BEHNCKE, "U.S. ARMY,"
Former Commander of Milwaukee Recruiting Battalion

"LTG Bostick's leadership style was fresh because he had no prior recruiting experience or bias. He learned quickly, asked a lot of questions, respectfully challenged the tribal wisdom, and had little time for the folklore that permeated the command. He led decisively, sometimes with only 85% of the solution. His risk taking made the difference between success and failure. A four-year journey that improved Army readiness through successful procurement of new Soldiers."

~FRANK SHAFFERY,
Director, Recruiting Operations Center,
U.S. Army Recruiting Command

"I led recruiting initiatives for six USAREC Commanding General's starting with Tom Bostick. As good as they all were, none approached the level of change agent that General Bostick demonstrated. His willingness to listen, personal drive and ability to motivate subordinates and superiors alike were instrumental in transforming the culture of the command and the Army's view of Recruiting."

~RICK AYER,
Director, Commanding General's Initiatives Group,
U.S. Army Recruiting Command

"Tom Bostick was a visionary leader as Chief of the Army Corps of Engineers. He was able to harness the power and resources of local communities and marry it to the capabilities of the private sector to get the 'undoable' done."

~FRED MEURER,
City Manager of Monterey, and the pioneer of public-private-partnerships with the military.

WINNING
after LOSING

WINNING
after LOSING

BUILDING RESILIENT TEAMS

Lt. Gen. (Ret.) Thomas P. Bostick, PhD

For Renee and Joshua

CONTENTS

PREFACE   **xv**

INTRODUCTION   **xxi**

CHAPTER 1   DISCOVERING OPTION 3   **1**

CHAPTER 2   THE MAILMAN DELIVERS . . . ONCE AGAIN   **17**

CHAPTER 3   WINNING IN SPORTS . . . WINNING ON
THE BATTLEFIELD   **27**

CHAPTER 4   FROM MISSION IMPOSSIBLE TO MISSION
ACCOMPLISHED   **41**

CHAPTER 5   THE PORT OF RIJEKA, TUZLA AIRFIELD,
AND KOSOVO   **51**

CHAPTER 6   ON THE POWER OF CALM AND QUIET
LEADERSHIP   **65**

CHAPTER 7   FIRST TEAM   **77**

CHAPTER 8   THE POWER OF PERSEVERANCE   **89**

CHAPTER 9   TURNING DIRT   **103**

CHAPTER 10   FIRST, BREAK ALL THE RULES   **117**

CHAPTER 11 OUR SOLDIERS   **143**

CHAPTER 12 PANCAKES-FOR-DINNER STRATEGY   **157**

CHAPTER 13 THE POWER OF TEAM DIVERSITY   **171**

CHAPTER 14 AT THE INTERSECTION OF LEADERSHIP
AND PERFORMANCE   **183**

CHAPTER 15 THE ART AND SCIENCE OF IDENTIFYING
FUTURE LEADERS   **201**

CHAPTER 16 CRISIS MANAGEMENT AND LEADERSHIP   **213**

CHAPTER 17 LEADERSHIP AND PEOPLE—A WINNING
COMBINATION   **237**

CHAPTER 18 PUBLIC-PRIVATE PARTNERSHIPS   **251**

CHAPTER 19 TALENT MANAGEMENT   **265**

CHAPTER 20 YOUR PEOPLE ARE YOUR BRAND'S BEST
AMBASSADORS   **275**

CHAPTER 21 FROM BOSS TO MENTOR TO FRIEND:
THE POWER OF RELATIONAL LEADERSHIP   **287**

AFTERWORD: WINNING AFTER LOSING IS ALL ABOUT
RESILIENCE   **297**

ACKNOWLEDGMENTS   **305**

NOTES   **317**

WHAT I LEARNED, CHAPTER BY CHAPTER   **337**

PHOTOS   **361**

ABOUT THE AUTHOR   **379**

A PERSONAL CLOSING NOTE   **383**

INDEX   **389**

PREFACE

I expect to pass through life but once. If therefore there can be any kindness I can show, or any good thing I can do to any fellow human being, let me do it now, and not defer or neglect it, as I shall not pass this way again.

~WILLIAM PENN

This book has been a long time in the making.

It represents the collected lessons of leadership I learned during my life of service in both the public and private sectors. It is a thank you to the many mentors who inspired and inscribed every step of my own journey, and I hope it will serve as a guide to those current and future leaders who are walking the leadership path today.

I am writing this book as a retired three-star General after thirty-eight years of service in the U.S. Army. Over the span of my military career, I served in many leadership roles, and each one taught me valuable lessons, whether I was serving in the trenches of tactical units, at the corporate level of The Pentagon, in complex public-private partnerships, or even in seemingly impossible global missions fraught with cultural and social challenges. From leading a platoon of just thirty soldiers to serving as Commanding General of the U.S. Army Corps of Engineers, the largest public engineering organization in the world, the lessons in this book are the distillations of the lessons I learned alongside others, the strategies our teams employed, the tactics we tested, the successes and even failures our teams experienced. Together we built winning, resilient teams. That is the Army part of this leadership book.

I am also writing this book as a former C-Suite leader in the private sector where I served as the Chief Operating Officer of a publicly traded bioengineering company with multiple biotech companies and R&D divisions. These companies and divisions focused on health, energy, the environment, and food with over one thousand employees—seven hundred with advanced degrees. There, the goals were different and yet very much the same. The skills in leading teams to achieve success in fulfilling their mission are similar whether it be in business development, in research, in new product development, or in mergers and acquisitions. There are lessons in this book that I learned in the private sector in a different economic environment with different resources. In business, teams must also become more resilient in order to win.

Based on my military as well as private sector experience, the lessons of leadership in this book apply to the public and private sector.

But there is an additional part of my personal leadership story that inspired me to write this book and sustained me through the months of research, writing, and re-writing. And that's my background. My personal history. My first tentative leadership steps.

Like so many others I grew up in a unique family. My mother was born and raised in Japan. My father was an African-American soldier from Brooklyn, New York. I have one sister and three brothers: five of us kids growing up as "Army Brats" moving from base to base, having to deal with change and learning to value every dollar. That is where my leadership lessons really began. My father, who was an athlete and champion runner, taught me the importance of self-discipline. Because there was not enough money to send five children to college, I learned to think creatively—and that's when I first considered the military as the path to a career. That is also when I was introduced to my first mentor—a man who showed me how to overcome seemingly insurmountable roadblocks and started me on a lifetime of mentorship—both being mentored and becoming a mentor. So, in several ways, many of my earliest lessons were foundational to both my successful careers in the military and in private industry.

Writers are always asked what inspired them to write. For me, the answer is simple. I have been teaching these lessons for a long time to small groups, large groups, and individual leaders. And each time I shared any of my personal lessons learned, the same request came up: "Wish you would write down all of these stories and lessons." And so, I have.

I began this book to share the lessons I learned in my own personal leadership journey to help other leaders. This book is not so much about specific situations, although there are many of those. It is more about how to reach down, dig deep, and find success when it seems almost impossible. My anecdotes and challenges are different from yours. But what remains the same are the challenges and goals of leadership—to build resilient teams, to forge strong organizations, to stretch for visionary goals, and to achieve mission success no matter what that mission may be.

The lessons I learned and share in this book are lessons that are universal yet the stories that forged them are unique. They are lessons that I hope will inspire you to try a new strategy, to set goals that are seemingly impossible, to take a risk, to never give up on a losing team, to use whatever tools come your way, even if they are not what you expected, and to listen not just to the voices from the top of the mountain, but to those voices shouting up from the bottom.

I began this book because I was asked to share what I learned not only when we succeeded, but when we did not. When I lost. And when my team lost. And when the mission failed. That is why I titled this book *Winning After Losing* because that is when I learned you need to really step up as a leader, and that is when winning is the sweetest. The theme of winning after losing is about resilience. How does a leader, organization, or nation prepare for an impact, absorb the impact, and bend but not break? It's about resilience.

As I write these words today, the world is a different place than it was a few months ago when I first put pen to paper. Today we are in the middle of a pandemic. The coronavirus has taken lives, scattered teams, shuttered businesses, and decimated global economies. It has also put a massive strain on the leadership skills of virtually every organization around the world whether private or public, large or small. Additionally, the tragic killing of George Floyd has caused the world to reflect more deeply on the fair and equitable treatment of African-American and other ethnic minorities.

Now I share lessons I learned over a lifetime of crises and challenges of every kind with soldiers, spouses, families, civilians, and some of our top leaders at the highest levels of government, including the White House.

I am honored more than ever to share these stories with you.

INTRODUCTION

Leaders think and talk about the solutions.
Followers think and talk about the problems.
~BRIAN TRACY

LEADERSHIP INNOVATIONS

How softball turned a losing team into a championship team; how a medical emergency triggered a powerful leadership lesson; how a major challenge led to the achievement of a seemingly impossible goal—these are just a few of the leadership challenges and innovative solutions presented in *Winning After Losing: Building Resilient Teams.* Based on true leadership challenges and their resolutions, the stories of success, achievement, and the overcoming of sometimes massive odds in this part textbook, part manual, part memoir, serve as compelling examples and lessons which at one time or another come to all leaders.

CRITICAL LEADERSHIP ISSUES

Leaders of organizations both large and small share similar issues: developing, nurturing, and honing the effectiveness of teams; setting goals that stretch abilities and yet are ultimately achievable; delegating; encouraging and balancing diversity and inclusion; judging when to follow prescribed policy and when to use individual initiative and take the risk to change; anticipating learning and training needs; managing globally dispersed teams across countries and cultures; keeping lines of communication open and flowing in multiple directions; and more.

LEADERSHIP SKILLS

Leaders of organizations also strive to optimize their leadership skills set in increasingly complex and fluid environments: communication across organizational levels; forging alliances and partnerships with new and different entities; balancing giving direction with allowing individual voices—sometimes the most inexperienced voices—to contribute to the conversation; planning across continents, social groups, cultures, and challenges; managing their own leadership education through rapidly evolving management changes; above all, reaching out to mentor and be mentored.

REAL NARRATIVES

Winning After Losing is a "turn-around" book. Frank, informative, instructional, and bold, it offers not only new leadership strategies but provides a fresh spin on the tried and true, the traditional approaches to leading teams and organizations. Every chapter provides the reader with a real leadership challenge and offers the real strategies and solutions that demonstrate the possibility of winning after losing.

WHO SHOULD READ THIS BOOK?

This book is a "must-read" for C-suite executives, organizational decision-makers, leaders in both the public and private sectors, entrepreneurs, and especially young "leaders-in-training" who are eager to learn critical management skills.

EVERY CHAPTER ILLUSTRATES WINNING LEADERSHIP STRATEGIES

In Chapter 1, *Discovering Option 3*, you will learn how an innovative softball game strategy turned a losing team into a championship team and gave the team members an unexpected bonus of leadership.

Chapter 2, *The Mailman Delivers . . . Once Again*, offers a glimpse into how to manage and optimize the unexpected—whenever it arrives.

Chapter 3, *Winning in Sports, Winning on the Battlefield*, models strategies based around competition, sports, and the respect for the history of the organization and how the lessons can be learned from the past and applied to the present. This chapter is about the leadership success when the goal is the team and its members.

In Chapter 4, *From Mission Impossible to Mission Accomplished,* you'll find the true story of how one leader's vision and care for his people flipped the impossible to possible. You'll also learn how you too can set a very challenging goal and know that your team will achieve it with the proper support and encouragement.

If you have ever been tasked with a seemingly unreachable goal, then Chapter 5, *The Port of Rijeka and Tuzla Airfield,* is your "must read" chapter. It contains two gripping true stories of how one leader set the bar high and inspired his team to reach the goals in record time. This chapter is both inspiring and instructive as the strategies described can be applied to your organization.

Chapter 6, *On the Power of Calm and Quiet Leadership,* deals with leadership skills that are often overlooked: the power of focus, calm leadership, and the belief that "people come first" no matter how many other seemingly "important" or "critical" or "priority" needs are in the way. This is the chapter that reminds leaders of what is important, what is really a priority, and who should always come first.

Chapter 7, *First Team,* is an important reminder that leaders must always maintain a delicate balance between individuals and the team or group they lead. This chapter offers three illustrations of just how to accomplish that balance in your own organization with your own people and teams.

Chapter 8, *The Power of Perseverance* was one of the most personally painful chapters to write and taught me one of my most important leadership lessons. It demonstrates how important it is that leaders take good care of their physical health so they can lead the teams and organizations that depend on their peak performance.

Chapter 9, *Turning Dirt*, outlines how a plan of action can be structured for leaders who rely on a globally dispersed workforce in the current global economic climate with all its cultural challenges.

Chapter 10, *First, Break All the Rules*, offers insight as to when to follow prescribed organizational policies and when to take that possibly career-risking move of breaking the rules.

Chapter 11, *Our Soldiers*, uses three compelling true stories to drive home the message that successful leaders inspire and motivate their people to achieve personal goals, and in doing so, strengthen the whole organization.

Chapter 12, *Pancakes for Dinner*, illustrates the power of remembering the often "invisible" members of teams—families and friends—who do not work for the organization directly, but whose support often drives success.

Chapter 13, *The Power of Team Diversity*, takes on the challenges of inclusion and diversity and offers inspiring examples of how to make these important challenges work successfully.

Leaders often come to a fork in the road to goal achievement and Chapter 14, *At the Intersection of Leadership and Performance*, is all about how to maintain that delicate balance between team leadership and team performance.

Looking ahead to future leaders who can pick up the baton is a key responsibility of present leadership. Chapter 15 is dedicated to strategies that define *The Art and Science of Identifying Future Leaders.*

Chapter 16, *Crisis Management and Leadership*, offers real and powerful examples of unique strategies to add to your leadership arsenal when dealing with a crisis management situation.

Chapter 17, *Leadership and People—A Winning Combination,* focuses on the skills required to select and manage human capital—one of the most important resources an organization has.

Organizations increasingly no longer work and thrive as independent entities but are connected locally, nationally, and globally through complex networks of partnerships and alliances. Chapter 18, *Public-Private-Partnerships*, outlines just how to set up and manage complex organizational partnerships.

Chapter 19, *Talent Management,* offers thoughts for current leaders, as well as emerging leaders, as to where to gain experience and skills in the organization and how to make the important career decisions among tactical, operational, and strategic leadership experiences.

Your People Are Your Brand's Best Ambassadors, Chapter 20, celebrates the importance of the people that make up a successful organization and the skill sets of leaders who put this powerful asset front and center.

Chapter 21, *From Boss to Mentor to Friend: The Power of Relational Leadership*, provides the key insights of relational leaders, which are so important during the COVID-19 pandemic. Even though relational leaders are focused on the goals and mission, they never lose sight of the people who make up their teams.

DETAILED LEADERSHIP LESSONS

In addition to the rich content, each chapter also contains a special section outlining specific lessons learned and offers suggestions for employing those self-same strategies in your own leadership role.

GET STARTED

Ready? Turn the page to start learning how softball turned a losing team into a championship team! It's the first of many stories detailing the strategies of leadership that allow teams to experience the immense satisfaction of winning after losing.

CHAPTER 1

DISCOVERING OPTION 3

A team is not a group of people who work together.
A team is a group of people who trust each other.
~SIMON SINEK

THE SOFTBALL CATALYST

This is the story of how a simple game of softball became the catalyst for one of the most powerful organizational team-building success stories of my career. What began as a last-ditch effort to mold a dysfunctional group of unmotivated, demoralized, uncaring team members, ended a few short months later with a turnaround team that was tight, trusting, cohesive, and had catapulted up the organizational leader board from last place to achieve the coveted status of first-place champions.

It all began with the MAIT. The acronym MAIT stands for the Maintenance and Assistance Inspection Team. MAIT inspections were serious business for the U.S. Army in Europe (USAREUR) in the 1980s. Even so, many companies regularly failed maintenance inspections from the V Corps MAIT team. But it was during my early years when I was a platoon leader assigned to the 54th Battalion in Wildflecken, Germany, a major U.S. training base not far from the former East German border, that I first understood just how important these MAIT inspections really were. I quickly learned that repeated MAIT failures could lead to a young Company Commander being relieved of a command, which in turn could damage a career.

It wasn't long after my time as a platoon leader that I was thrown into the MAIT challenge head-first and had to face the very real challenges of the MAIT failure problem.

THE CHALLENGE OF A FAILURE PROBLEM

My new assignment after serving as Platoon Leader was the position of Battalion Maintenance Officer (BMO). I had been learning a lot about maintenance. Not enough to pass the inspections but enough to begin to understand why we were failing. I worked for a superb Battalion Commander, who later went on to become a Lieutenant General and Superintendent of the United States Military Academy at West Point. My fellow officers and I had felt so fortunate to have one of the best officers in the Army leading our battalion. So, when the Battalion Commander asked me why we were having such a difficult time with maintenance inspections, I was ready to venture a preliminary analysis. I put forth the observation that as the Lieutenant charged with the responsibility of supervising maintenance, I really didn't understand maintenance very well and that I wasn't alone in this lack of understanding. I explained that most of our first-line supervisors, our Lieutenants, did not know how to supervise maintenance. To make matters worse, the Lieutenants had never been trained to supervise maintenance, and they didn't know what they were doing. As a result, they didn't have the tools to help train or supervise others.

The outcome of this meeting was the formation of a new "Saturday Certification Program." We put together a supervisory certification checklist, and every Saturday the Battalion Commander personally certified each Lieutenant, one-by-one, until they were all certified to supervise every piece of equipment under their responsibility. Not only was this a learning opportunity for our team of Lieutenants, but it was also a learning opportunity for me.

As the BMO, I learned how to maintain all equipment from our trucks to our communications equipment, from our generators to our trailers. After about eighteen months as the BMO, I was assigned as the B (Bravo) Company Commander. This assignment was a special opportunity, and I was determined to succeed. What I didn't realize at the time was that not only would I succeed beyond any personal expectations, but that's where the "Softball Strategy" for team-building would be born—a strategy that would result in a group of soldiers going from losers to champions—a winning team based largely on the trust that they gained in each other.

At B Company, our maintenance situation was abysmal. We repeatedly failed the critical MAIT inspections. It was no comfort to know that we weren't the only company failing. It was clear that if we did not turn things around soon, my time as a Company Commander would be short-lived.

THE BLAME GAME

Among our soldiers, everyone was playing the blame game. The drivers blamed the repair parts clerks. The repair parts clerks blamed the mechanics, the mechanics blamed the supervisors who did not know how to supervise. There was a lot of finger-pointing and a general lack of trust.

We had to build a team that would work together, trust each other, and leverage the skills of everyone on the team toward a single goal—passing the MAIT.

The solution? Softball. If we couldn't build a winning organizational team, a softball team seemed like a good method for building a team outside of the work environment. We had to build a team that trusted in each other. Our softball team was a true pick-up team with very few skills as a team and little superstar talent. We thought the softball field would allow us to enjoy each other's company, to socialize, and to laugh together. We also thought that these bonding activities could serve as foundational pillars of trust. We were determined to build a functioning, cohesive team that would ultimately transfer the concept of teamwork from the softball field to our company maintenance program.

Our first effort did not result in the win we had hoped for but in another total failure. It was all because of the beer.

BEER BALL

The troops wanted to play "beer ball." I wasn't familiar with the game, but it didn't take me long to catch on. We were stationed in Wildflecken which is situated in the Rhön Mountains. Atop one of the mountains is an area known as the Kreuzberg, home to a famous old monastery founded in 1731 whose Franciscan monks brewed some of the best beer in the world. The beer was not bottled so if you wanted to sample it, you had to visit the monastery. If you wanted to bring some beer back down the mountain, you could buy a keg of the fantastic brew. And so "beer ball" became our first softball game. Members of our team set up the rules, and the rules were simple. There was a keg of beer at every base. When you arrived at a base, you had to drink a cup of the beer while waiting to advance to the next base. Our skills and our softball game hit a new low. I decided that was our first and last time playing beer ball.

That's when we decided to get serious and started playing to win. We decided to build a better softball team. That decision was the first step to turning everything around—from softball to those all-important MAIT inspections.

THE BUS STRATEGY

Meanwhile, we were faced with another big challenge. At that time in the Army, there was a serious problem with drugs and alcohol. Some troops spent their money getting high and not taking advantage of the remarkable opportunity of exploring Europe, which was easy to do from our base in Germany.

My wife Renee, my First Sergeant, and I were determined to find a way to make the soldiers turn their focus to the enjoyment of living in Europe. We decided that we needed a trip strategy for visiting Europe. As it turned out, those trips played an integral part in our Turnaround Softball Strategy.

Our first trip would be very inexpensive. We would use a military bus, at no cost, and link the trip to leader development training. I asked the soldiers where they wanted to go, and a few said they wanted to go to the "Wall" in Nuremberg, Germany. I dutifully filled out the trip request for leadership training. It wasn't long before my Battalion Commander called me to his office to ask about the planned trip. His first question was whether I knew what the "Wall" meant? I told him I understood that Nuremberg was a walled city where they had held the Nuremberg trials after the Second World War. It had seemed like an excellent leadership development opportunity to me. My Battalion Commander explained that the "Wall," for the soldiers, had nothing to do with the part of history that I referred to—it was the name given to the red-light district in Nuremberg. Embarrassed, I agreed to take the troops to another, less racy part of Nuremberg, and the trip was approved.

On the Saturday morning of the trip to Nuremberg, there were exactly four people on the bus: my wife, my First Sergeant, one Lieutenant, and me. Not a single young soldier had shown up. Together my First Sergeant, the Lieutenant, and I went to the barracks to look for troops who were not too hungover from the previous night. We put the ones we found on the bus going to Nuremberg. It was mandatory fun for these few soldiers. Even though the trip got off to a rocky start, everyone ended up having a great time. It was such a success that the few soldiers we had gathered up spread the word. It wasn't long before we had requests from the soldiers for another trip.

Our next trip was much better. In the early days of my career, soldiers received their pay in cash from one of the Lieutenants serving as the monthly pay officer. This was a duty that not many Lieutenants looked forward to because there was always the chance that they would accidentally dole out more cash than required. Then the pay officer had to make up the difference from their personal funds. But to help the soldiers pay for the increasingly popular trips, each pay period we set up a table next to the pay officer where we collected enough cash from the troops to pay for their food, travel, hotel, and any small additional expenses. Our next destination was Lloret de Mar, Spain. What a difference from our first venture to Nuremberg. This time the bus was filled to capacity with troops, and for many it was the very first time they had traveled anywhere in Europe outside Germany. The trip to Spain was a huge success.

But what did these trips have to do with softball? And more importantly, how did these trips and softball become the catalyst for creating a powerful success strategy dominated by cohesive, trusting teams?

We were about to find out.

THE BIRTH OF A CHAMPIONSHIP TEAM

Our softball team was starting to improve. We learned how to match the skills and abilities of each soldier to the position on the team where they could add the most value. Even though we had little hope of distinguishing ourselves, we entered the Wildflecken softball championship. Our teamwork paid off. We didn't win first place, but we did win second place in the Wildflecken Softball Championship. Not bad for a team that had launched their softball career with kegs of beer.

The top two teams from each military garrison were invited to a four-day U.S. Army Europe (USAREUR) Softball Championship tournament in Hohenfels, Germany. Our second place standing at Wildflecken secured us a spot in the USAREUR Championship tournament. Once again, our pickup team from Wildflecken was not expecting much. We were up against some outstanding teams. We weren't surprised when we lost our first game to a field artillery team. But it was how we lost that was critical to our future "Softball

Strategy" of team excellence. The game was ended early due to the "Ten-Run Rule." We were losing by ten runs and in softball that's crushing, so the umpire invoked the Ten-Run Rule to end the game early and put us out of our misery. Here we were one game into the tournament and already back to being losers. But we told ourselves that it was great that we had made it this far. At this point, we still had no idea of just how far we would go thanks in large part to that Ten-Run Rule.

The next day was completely different for the losers. We won! And then we won again! We kept winning! After two days of softball, we found ourselves in the championship round. And in a pure stroke of luck we were to play against that very first team, the artillery unit, that had beaten us by the Ten-Run Rule! To win the tournament, since we were coming out of the loser's bracket, we had to win twice against the artillery unit.

We played the first game. It was tight. But thankfully, we won.

Now it was time to play the second and deciding game. But there was one challenging issue. Later that same day, we had another one of our successful "Explore Europe" trips scheduled, booked, and paid for by many B Company soldiers. Two buses would be departing from Wildflecken in front of Bravo Company Barracks to take the troops to Rome. And almost all the soldiers on the softball team had already paid and were scheduled to go on that Rome trip. The buses would be leaving while the championship game was in full swing—likely before the game ended.

I pulled the team together and told them we had two options. We could either play the game for the championship and miss the bus for the trip to Rome, or we could get on that bus with our second-place trophy and forfeit the championship game. I told them that if they decided to stay and play for the championship, I couldn't return any of their trip money to them. I told the troops how proud I was of them. No one expected us to come this far. I told the team that I would still be very proud of them if they decided to leave the tournament now with the second place trophy and go to Rome. I told them that no matter which one of these two options they chose, no matter what decision the team made, I would support it either way, and I would be very proud of their achievement.

The team went into a huddle. I stood to the side and waited.

THE UNEXPECTED OPTION 3

It was one of the team leaders, a Sergeant, who finally walked over to me and said, "Sir, we have made our decision." I asked him if the team had picked Option 1, to stay and play for the championship or Option 2, to depart now with the second-place trophy, and in time to make it on the bus to Rome. He surprised me by saying, "Sir, we came up with an Option 3." I reminded him that I hadn't offered an Option 3. But he insisted. "Sir, we understand that, but you know how the field artillery team ended our first game early by invoking the Ten-Run Rule? Well, we're going to do the same thing to them. We are going to win that championship game by the Ten-Run Rule, force an early end to the game, and still have plenty of time to spare to make it back to Bravo Company and catch that bus to Rome."

Listening to my Sergeant discussing Option 3, my eyes welled up with tears. At that moment I knew we had created a team with the spirit and the will to win. I realized that I had boxed them in with only two options, but they were not willing to accept that. They pushed beyond the constraints of two options. They were creative. They acted as a team. And what did I learn? I learned that great ideas could come from anyone on a team if the environment is open to communication by all, safe, and trusting to allow for this type of freedom of thought. I also learned that as a leader, I should never box a team in with only two options.

As we went into that final game, the tension was high, but the motivation of the Bravo Company troops was even higher. They scored run after run after run until finally Bravo Company ended the game early by defeating the artillery team by the Ten-Run Rule, just as they had said they would, and won the 1982 U.S. Army Europe Softball Championship! It was truly an amazing accomplishment.

As we pulled into the Bravo Company parking lot, the buses with engines running were waiting to take the U.S. Army Europe championship softball team to Rome.

The team enjoyed two victories that day. A softball championship and a trip to Rome. But I knew that through their own imagination, perseverance, and teamwork they had won a third victory—an Option 3 victory. They were winners. They were champions. They were a team!

But the story doesn't end there.

During that year, the skills our team developed on the softball field and the camaraderie that bonded them during our many travels through Europe started to pay off in our maintenance program—the ultimate challenge that had started this whole journey to team-building in the first place. With the superb leadership of my Executive Officer and Motor Sergeant, combined with the new spirit of a winning team, Bravo Company started to pass all its inspections with flying colors. We were soon nominated to compete in the best maintenance company competition in U.S. Army Europe, and we won! That was just the beginning. We entered the incredibly challenging competition for the best maintenance company in the entire U.S. Army. It was not easy. We trained intensely and worked as a team just as we had in our softball competitions. Inspectors from the United States visited our company and inspected every facet of our program. Then, in 1983, Bravo Company, which started out with one of the worst maintenance programs, was named the best mid-sized maintenance program in the entire U.S. Army. We had done it. We were number one. We were champions!

After we won the best maintenance program award, people would ask me about our secret to success. How had we turned our losing team into a winning one? And I would always respond, "We played softball and traveled on a bus all over Europe."

I learned many life lessons as a young Company Commander. I learned that building a winning team is always possible, even without a bunch of A players. I learned a leader can change the behaviors and attitudes of the members of a team and a team can change the thinking of its leader. I learned a team can develop the skills to face adversity, trust each other, and win. I learned that these winning outcomes could happen when leaders show they care for their troops. And I learned never to box a team into a fixed set of options. I learned that there is always an Option 3.

WHAT I LEARNED

1. **Allow your team to offer alternative options**. Having different options is a recognized leadership skill; however, what I learned was to let your team come up with some options of their own and implement those options as alternatives. You may find that sometimes the best path to a winning outcome may not be the most direct. Softball and our trips inspired teamwork, trust, and a winning spirit. This idea of a team-developed "Option 3"

is a leadership strategy that I have since used frequently to good effect.

2. **No need for all A players to build a championship team**. All too often leaders feel they can't build outstanding performance-driven teams unless they have a team made up of only star players, A players. I found this not to be the case. We had few A players, and even so, managed to build a championship team in both sports and maintenance. A big part of winning for any team is teamwork and most of all trust.

3. **Trust**. Even if a team is made up of a seemingly random mix of people with very little in common, they can bond strongly and develop interpersonal trust.

4. **Let individual talents shine**. Even though a team may be focused on one single goal, each member of that team has different skills to contribute. A team does best when each member is placed in a position where their individual skills and talents shine. Jim Collins highlighted a similar concept in his book, *Good to Great: Why Some Companies Make the Leap and Other's Don't,* in which he stressed the importance of getting the right people on the bus, then getting these people in the right seat on the bus, before trying to figure out where the bus is going. Although *Good to Great* was published well after we developed a championship softball and maintenance team, we used the same concept as the bus analogy.

5. **Downtime**. Many of us work in high-performance settings as do our teams. But it is important to keep in mind that a

team needs to have some downtime or a non-work-related activity (in our case it was softball and traveling Europe) to help bring the team together.

You have no choices about how you lose, but you do have a choice about how you come back and prepare to win again.

~PAT RILEY

CHAPTER 2

THE MAILMAN DELIVERS . . .
ONCE AGAIN

No Mission Too Difficult.
No Sacrifice Too Great.
Duty First!

~MOTTO, 1ST INFANTRY DIVISION, U.S. ARMY

WHERE YOU LEAST EXPECT IT

Sometimes a team needs a little help. And sometimes that help comes from a place where you least expect it. That's what happened to the soldiers of the 1st Engineer Battalion during one of their most dangerous missions ever—fighting the raging wildfires in Idaho. They needed a little help and they got it—from "The Mailman," Karl Malone, the NBA Basketball Star.

It was 1994. I had the honor and opportunity to command the 1st Engineer Battalion for two years, including a period of time in which we were deployed near Boise City, Idaho, to support the U.S. Forest Service in fighting massive fires that had spread over eight thousand feet up the mountains.

This was not our usual fight. This was not our usual battle. Army units prepare for war through challenging training at the Fort Irwin National Training Center (NTC) in the Mojave Desert. But while conducting preliminary leader training in preparation for our battalion's deployment to the National Training Center, I received a call from my Division Commander, who directed me to return immediately from the National Training Center to Ft. Riley and prepare to deploy my battalion, along with a chemical company, within the next forty-eight hours to fight the 1994 wild fires in Idaho. With only forty-eight hours to prepare for a mission that was completely foreign to our soldiers, we focused closely on the training received from the fire-fighting experts from the U.S. Forest Service.

Even though this fight was not our usual fight, our battalion was well equipped, not only by training but also by history, to fight and win whatever the battle. The history of this battalion dates to May 15, 1846, when a company of Miners, Sappers, and Pontoniers was formed at West Point, New York, on the banks of the Hudson River. Alpha Company, of the 1st Engineer Battalion, was a direct descendant of that company at West Point. The long and storied history of the battalion has served as a rallying point for this battalion which remains the most decorated engineer battalion in the Army.

And it was this battalion that would receive a very special delivery from The Mailman to add to its already extraordinary history.

We arrived near Boise City, Idaho, with all our equipment and prepared to take on the mission of fighting fires. But this task would be different from anything that we had done before. As we looked up into the sky, and up the steep slope of the mountains, we saw the nearly eight thousand feet of elevation almost obliterated by a massive smoke cloud covering the skyline. We needed specialized training, and we needed it fast. And we received it. Our directions and daily mission assignments came directly from the U.S. Forest Service.

The first order of business was to move our task force to our basecamp much higher up on the mountains and closer to the fire. Our task force included approximately five hundred total soldiers, including all of the 1st Engineer Battalion in addition to a chemical company. The Forest Service directed that we would put our equipment on trucks while our soldiers would travel by bus to the basecamp. I pointed out that we do not separate our soldiers from their gear. The Forest Service insisted on this guidance, and they were in command. Our equipment and soldiers would travel separately, but the Forest Service assured me that the two would link up at the basecamp.

Despite the August summer heat, when we arrived at basecamp later that evening, the temperature registered a freezing 15 degrees. It was cold. We were all shivering. And not surprisingly for me, our equipment never made the promised link up. We had no warmer clothing, no sleeping bags, nothing. Fortunately, we were able to secure some extra blankets from the fire-fighting teams already in place, so that the soldiers could sleep that first night. Our equipment arrived the next day—a sight for sore eyes.

We immediately went to work on the firefighting mission, but it didn't take long for us to realize that we had several difficult challenges. First, while we were in great physical shape, we had been training in desert conditions where the land was mostly flat. We had not trained in mountainous terrain and both the steep mountain slopes and the high elevation pushed our physical capabilities to their limit. Second, even though we finally had our own equipment, we quickly realized that it wasn't effective—especially our boots. The ground was so hot that the heels were melting right off our boots. We had to order new fire-resistant boots. And finally, the conditions themselves were life-threatening. We fought fires during the day and night. But the night firefighting proved to be extremely dangerous. Due to darkness, we couldn't see falling trees until we heard their thundering whoosh sound as they smashed into the scorched ground close to the troops. After a few very close calls, I decided to stop night firefighting.

A PROBLEM AND MAYBE A SOLUTION

After a few weeks, our soldiers became more comfortable with the daily task of firefighting, settled into a routine, and began to look for things to do during their brief time off. Sports became their free time focus. However, there was a problem. Since we had to deploy so quickly on the fire-fighting mission, and because the mission was so new to us, we had not even considered bringing any sports equipment with us. And now the troops who were acclimated to the new environment felt stuck on a mountain with nothing to do during their downtime.

Luckily, there was a town not too far away. A few of our soldiers asked if they could go into town and purchase some sports equipment. I thought it was a great idea. I gave them a little cash and reminded them that this was a one-day-only mission. Their orders were to be back by the end of the day.

WHEN ONE DOOR CLOSES . . .

The soldiers arrived in town later in the day than they had expected. When they finally located the sporting goods store it was closing. The soldiers explained to the store manager, who was locking the door, that they had just come down from the mountains where they were fighting fires and wanted to get a few sports items. The manager would not budge, saying that they could just come back tomorrow. The soldiers tried again. The troops explained that they couldn't return tomorrow, that their orders were to return to their base camp tonight, and that this visit to town was their only opportunity. But the manager was unmoved by the soldiers, by the fires they were putting out to protect the town and the store, by their simple request, and just firmly repeated that the store was closed and locked the door.

ANOTHER DOOR OPENS

Disappointed, the soldiers decided to make the best of it and walked around exploring the town. Soon they found themselves at the local mall. Inside, there was a large group of people all crowded around a table. They were waiting for their turn to get an autograph. The soldiers came a little closer. There, sitting at the table signing autographs, was none other than basketball great, Karl Malone, considered one of the greatest power forwards in NBA history, who was nicknamed The Mailman since the time he played basketball at Louisiana Tech University because he "always delivered."

Malone looked up, saw the soldiers, and asked what they were doing in town. They told him they were fighting fires and had come down the mountain to purchase some sporting equipment to use when they were off duty. Malone told them that there was a good sports store in town. The soldiers thanked him and explained that they had already been to that sports store but that the manager was closing and wouldn't let them in to buy anything—that the manager told them to come back tomorrow. They told The Mailman that they only had this one day and had to get back up the mountain to fight those fires again before tomorrow. The soldiers explained that there was no tomorrow for them to buy sporting goods.

Malone stepped away from the table, assuring the crowd that he would be back. Then he took the soldiers back to that closed and locked sports store. Karl Malone knocked on the door and when the manager appeared behind the glass door,

The Mailman said, "Hi, I'm Karl Malone, do you mind if I do some shopping?" The store owner didn't hesitate but unlocked the door, smiling, and said, "Absolutely, please step right in, Mr. Malone." The Mailman stepped inside followed closely by the two soldiers—the same soldiers who had been turned away from that store just a short time before.

The soldiers purchased frisbees and footballs for the rest of the team up the mountain. The Mailman threw in a bunch of basketballs and autographed them for the men. Then he paid for everything in cash. The store manager personally escorted the group to the door, and just as the door was closing, The Mailman held the door open. He looked at the store manager and said, "I know you only opened the store after closing because of who I am, but the real heroes are these soldiers fighting those fires up on the mountain." Malone said that, in the future, he would appreciate it if the store owner would treat the soldiers the same way he was treated. Karl "The Mailman" Malone really delivered that day for our soldiers and our task force. The soldiers returned to the basecamp and the battalion immediately began enjoying the sports equipment. After the 1st Engineer Battalion retured to Fort Riley, the autographed basketballs were placed in the trophy cabinet as a reminder of the great support from "The Mailman."

Years later I had another opportunity to see The Mailman. Each year the U.S. Army hosted the U.S. Army High School All-American Bowl Football Game. Years after my time as the Commander of the 1st Engineer Battalion, when I had become a two-star General and Commander of U.S. Army Recruiting Command, I invited Karl "The Mailman" Malone to one of the games. He was a big hit and still very supportive of the Army team.

Our experience in Idaho taught us some great lessons. We learned that we could win battles on any terrain and against any enemy. We learned that a team was important to achieving any goal, whether on the battlefield or the side of a mountain. We learned that a bunch of frisbees, footballs, baseballs, or basketballs were important in reinforcing the value of teamwork by building up the qualities and attributes that help make a group of soldiers successful. We continued to focus on winning in sports and motivating our team through "fun" activities where they could excel and build confidence, trust, teamwork, and the leadership qualities that would enable them to win in many other areas. But we also learned one very valuable lesson—that sometimes help is "delivered" by a very unexpected mailman and that we could all be those mailmen who deliver for others!

WHAT I LEARNED

1. **Be agile and open.** Being agile and open means being prepared through creativity and a positive mindset to accept help from unexpected sources.

2. **Great training has universal applications.** A well-trained team can win battles on any terrain and against any

enemy. Similarly, well-trained teams of civilians can adapt and win in a completely different area with the proper training, resources, teamwork, and leadership.

3. **Don't dismiss the unconventional**. Unconventional tools like the footballs and frisbees that our team used may be the very glue that binds a team together.

4. **Appreciate the power of sports**. Sports play a large part in forging and strengthening many different types of teams in more cohesive and effective units.

Those who are happiest are those who do the most for others.
~BOOKER T. WASHINGTON

CHAPTER 3

WINNING IN SPORTS
WINNING ON THE BATTLEFIELD

Competing at the highest level is not about winning.
It's about preparation, courage, understanding, and
nurturing your people and heart. Winning is the result.

~JOE TORRE

Organizations need both leadership and teamwork. Without effective leadership, inspiring leadership, and committed leadership, teams lack focus, direction, and confidence. They are not effective. And when teams are not effective, the entire organization suffers. Leaders and their teams are inextricably linked—forged together for strength. These are three stories where the leadership and team building had impressive outcomes.

#1 THE STORY OF THE SPORTS AWARD

Sports are a powerful team-building tool and if used correctly have the potential to not only help achieve team-specific goals, but go far beyond to build confidence, inspiration, and achievement. That's what happened to my battalion, the 1st Engineer Battalion, 1st Infantry Division at Fort Riley, Kansas.

Our battalion was training in the field for their upcoming rotation at the National Training Center. The National Training Center is where some of the Army's most important and intense training takes place. It was tough and demanding training, and units prepared long and hard just to get ready for their time at the National Training Center. In the middle of all this preparation, our Battalion Command Sergeant Major approached me and told me that the Fort Riley wrestling championship was the next day. We had to make a quick decision. Should we leave the troops in the field training for their upcoming all-important National Training Center rotation, or should we pull our best wrestlers out, send them into the wrestling competition, and then bring them back to the field with the other soldiers? Winning is contagious whether on the battlefield or on sports field. Decision made. I told the Battalion Command Sergeant Major to send the best wrestlers to the wrestling championship and then bring them back to the field exercise.

Why did we interrupt their training? Because sports can be the glue that often binds a team together. Sharing victory in sports as well as sharing defeat in sports—both make a team strong. And that strength translates to their performance on the field of battle.

The wrestling championship was only one example where training in the field was interrupted for training in the athletic arena. And it always paid off. Through the many sacrifices and the creativity of finding ways to allow our soldiers to compete while we were in the field, our battalion finished first in cross country and several other competitions, and we ultimately won the 1st Infantry Division award of the Commander's Cup Trophy for achievements in sports during the year! After leading a battalion run, our Division Commander presented us with the Commander's Cup Trophy. One of our soldiers who trained for the boxing tournament said it best: "We are allowed to practice and have an opportunity to participate in any sport. Winning the Commander's Cup shows our battalion accomplished something together. It's something to be proud of."

As American Olympic champion Kristi Yamaguchi observes: "An athlete gains so much knowledge by just participating in a sport. Focus, discipline, hard work, goal setting and, of course, the thrill of finally achieving your goals. These are all lessons in life." I have always believed that everyone wants to be on a winning team. Providing members of any team with the opportunity to win is critical for all leaders. Team members may or may not win, but having a shot at winning is very important.

#2 THE STORY OF ARTILLERY FIRE THAT DID NOT HAPPEN . . . AND THEN IT DID

Winning in sports gave many of our soldiers the qualities that would allow them to win on the battlefield. They developed persistence, endurance, creativity, and above all the passion to win.

It wasn't long before we had a chance to test our theory—that sports make for better teams. It was during one of our opportunities at the National Training Center that our team, the 1st Engineer Battalion, demonstrated it knew how to win.

But it didn't begin very well. In fact, it was actually a disaster.

At the National Training Center, visiting units train by fighting against the Opposing Force (OPFOR). Now it is important to understand that the Opposing Force generally defeats every brigade (a unit of about five thousand soldiers) that arrives at the National Training Center. Why? Simple. First, the OPFOR knows the terrain better than the visiting units. Second, they fight a visiting brigade almost every month, so they are highly trained. Third, the OPFOR is a highly cohesive team. Fighting and defeating the U.S. Brigades is the OPFOR's primary mission and they are excellent at accomplishing that mission. The visiting units come to the National Training Center perhaps every twelve to eighteen months so the OPFOR receives much more training than any visiting unit. The bottom line is that very few units ever defeat the OPFOR.

And at first, it seemed like we would also join the ranks of defeated units. But through a combination of leadership and teamwork, the 1st Brigade, supported by the 1st Engineer Battalion, found a way to win again and again.

Initially, it didn't look good for our unit. Our biggest challenge appeared to be lack of training in the National Training Center environment, especially in battle-specific skills like calling for fire—artillery support. For example, one of the Battalion Commander's units was being attacked. The Battalion Commander requested artillery support several times. But his unit was not the priority, and they received no artillery support. Finally, the priority shifted to this Commander. Now the Battalion Commander was first in line, and he was directed to call for artillery fire. But the situation quickly went from bad to worse. As we all listened in on the brigade command radio frequency, we realized that the Battalion Commander failed to give a direction—a crucial component—in calling for the artillery fire. At that moment, I realized that we might have a big gap in our own battalion training. I didn't know if my leaders could correctly call for fire, for artillery support, while under the pressures of combat. And while that was a serious problem during a training exercise, in an actual battle it could be fatal. The unit of the Battalion Commander that did not properly call for fire was decimated, but thankfully, this was training.

When we returned to Fort Riley, Kansas, I was determined that every officer in the battalion would become certified to call for artillery fire support. We would never be exposed and vulnerable again. From my days as a Battalion Maintenance Officer in Wildflecken, Germany, I knew how important it was to train and certify leaders to accomplish their important duties.

We went to work. We went to the Field Artillery Simulation Center that replicated the National Training Center. This would eliminate any excuse that we weren't familiar with the environment, terrain, or any other factor. We planned to first train our Lieutenants who would be forward in the field and in the best position to call for fire. Our trainer called out, "Three enemy tanks in the open . . . call for fire . . . *Lieutenant* Bostick." That was me. Although I was a Lieutenant Colonel, the trainer called me "Lieutenant" so I could set an example for all the other Lieutenants. With the assistance of some preliminary training, I did it. I successfully called for artillery fire during the simulated battle. Leaders must lead by example. If something is difficult for others to do, the leader should be the first to master the task. That's what the trainers did with me as the example, and then we trained every officer to do the same.

It wasn't long before we had a chance to show our newly acquired skills with a surprising and very satisfying result.

It was at our next National Training Center rotation. Our Brigade Commander made a bold move. He moved the 1st Engineer Battalion north of a mountain range called the Granite Mountains. That's where we would conduct an operation that we had never conducted before in training. The

1st Engineer Battalion would conduct a movement to contact and then defend the Alpha-Bravo pass while also setting up a hasty defense for the rest of the brigade south of the Granite Mountain. A movement to contact requires a unit to advance until it makes contact with the enemy. The operation involved two armored and one infantry task force of the brigade simultaneously conducting a movement to contact and then a hastily prepared defense south of the Granite Mountain range in the Central Corridor. The brigade was obviously taking a huge risk in the north in Alpha-Bravo pass.

During the preparation for the upcoming battle, our Division Commander spoke on the Brigade Command radio frequency asking for Dagger 6, my Brigade Commander, and Diehard 6, me, to meet him at the "Iron Triangle," which is a well-known meeting point in the Central Corridor of the National Training Center.

The Division Commander then asked Dagger 6 to describe his plan. After listening to the plan, the Division Commander said, "It sounds like you're sending Bostick up to Alpha-Bravo pass with a shotgun and three shells." The Division Commander went on to comment that if he were the Opposing Force, he would send the entire OPFOR regiment north and overwhelm the engineers in Alpha-Bravo pass.

My Brigade Commander reassigned some of the brigade's resources and created a task force in the north, called Task Force Bostick. Task Force Bostick consisted of our engineers, an entire tank company, and a Bradley platoon of infantry under my command. Not only that but the task force had priority on artillery and air support. We worked all evening planning for battle. Typically, engineer units use inert mines and concertina to block Alpha-Bravo pass. Our battalion changed up the typical strategy and elected to use dozers to cut an actual ditch in Alpha-Bravo pass that would be impenetrable without a bridge crossing vehicle. My Battalion Operations Officer and I would work with the troops through the night conducting a hasty defense in the Central Corridor as well as in Alpha-Bravo pass.

In the next part of the plan, I instructed one of the fastest runners in our company, who was nicknamed "Roadrunner," to move well out in front of Alpha-Bravo pass to the point where the OPFOR had to make a decision to go north or south of the Granite Mountains. That critical decision point was ten kilometers in front of our position. Roadrunner then moved quickly on his own as a forward scout. We then positioned another scout forward and at a high point. We let him know that should he identify the OPFOR Combined Arms Reserve, the element that would close with and destroy the enemy, he should immediately call for artillery fire on the brigade command frequency.

The battle began. It wasn't long before Roadrunner called, saying that he saw a large cloud of dust heading for the Granite pass, moving north. We immediately asked if it was the entire regiment. He said it was. Our Division Commander was correct regarding the entire OPFOR regiment attacking north to overwhelm Task Force Bostick. Our Lieutenant scout spotted the Combined Arms Reserve, quickly entered the brigade command radio frequency, and successfully called for fire. His fast action would later earn him an impact Army Achievement Medal.

In the meantime, the remainder of the entire OPFOR regiment moved north of the Granite Mountains heading directly for Alpha-Bravo pass. Unfortunately for them, they did not lead with their assault bridges and as a result were forced to stop completely in the Alpha-Bravo pass.

We called for artillery as well as air support into the Alpha-Bravo pass and decimated those vehicles in the pass.

Unsuccessful in penetrating Alpha-Bravo pass, the remaining OPFOR vehicles moved one by one over the Granite Mountains along a small winding road that exited into the Central Corridor. And that's exactly where the three task forces of the 1st Brigade were in position waiting.

In that battle, the 1st Brigade and Task Force Bostick had accomplished something that had not been accomplished before. The 1st Brigade created an Engineer task force, Task Force Bostick, and every OPFOR vehicle was killed, something very rare for the OPFOR. We had created an engineer task force that successfully executed a mission for which it had never trained, and ultimately the brigade destroyed the entire OPFOR regiment with support from every part of the brigade including highly trained and confident engineer Lieutenants.

That was a win we could all be proud of—leaders and team members alike. It was a win that involved lots of hard training, imagination, and tight teamwork. And it worked!

#3 THE STORY OF THE MINEFIELDS

Who would have ever thought that a battle that took place in the middle of the twentieth century would serve as inspiration for the team in the twenty-first century? But that's exactly what happened.

During the Normandy landings at Omaha Beach in 1944, the 1st Engineer Battalion led the assault forces by opening gaps in the enemy mine and wire obstacles. From December 16 to January 25, 1945, the Battle of the Bulge, which Winston Churchill later called the "greatest American battle of the war," raged in Europe. The 1st Engineer Battalion played an important role in that historic battle by placing over thirty thousand mines in the snow in just ten days. That was an average of placing three thousand mines a day.

That feat inspired the 1st Engineer Battalion in another battle they fought during their National Training Center rotation during which the brigade had to take up a deliberate defensive position. One of the key tasks during defensive operations was to lay minefields. Some of these minefields were inert and some were live, real mines. Typically, engineer units would place several hundred mines during a deliberate defense. Other, better-trained units might be able to lay over a thousand mines. Before deploying to the NTC, the 1st Engineer

Battalion reviewed its history during WWII. They revisited the feat of their historical brothers of laying thirty thousand mines in ten days, in the snow. That was an average of three thousand mines a day. Our unit became determined to honor the World War II heroes by placing the same number—three thousand mines during the one-day deliberate defensive battle at the NTC. That number three thousand represented the total number of mines at NTC.

As the preparation for the defense continued, I visited the live (non-inert) minefield where a Sergeant was rallying the troops. Each time I stopped by the minefield site, this young and highly motivated Non-Commissioned Officer asked me, "Sir, how close are we to three thousand mines?" The Non-Commissioned Officer and all the troops wanted to achieve the goal that we set out for ourselves. The battalion placed all the mines available at the NTC—over three thousand—and the 1st Brigade once again killed every OPFOR vehicle in the regiment in this battle.

Having witnessed the importance of history in motivating our troops, I reached out to a past Commander of the battalion for assistance with our upcoming Annual Engineer Ball. Lieutenant Colonel William B. Gara commanded the 1st Engineer Battalion from North Africa to the 1944 Normandy Landing and through the Battle of the Bulge in May 1945. Bill Gara agreed to be our guest speaker. He also conducted professional development discussions with our officers, visited with our troops, promoted and re-enlisted members of the battalion, and gave a memorable presentation as our guest speaker at our Annual Engineer Ball. Each time that I walked down the long corridor to my office, I gazed at the past Commanders of the 1st Engineer Battalion. I always stopped at the photo of Lieutenant Colonel Gara for two reasons. First, he looked very young. Secondly, he commanded during one of the most challenging times during the Second World War for the 1st Engineer Battalion. When I finally met Colonel Gara, I asked him how old he was during the war. He thought, then said, "Let's see, I started to command in North Africa, landed in Normandy in 1944, and then led the battalion through the Battle of the Bulge in May 1945." Then he looked at me and said, "When I landed in Normandy, I had over 1,400 troops under my command, and I was twenty-seven years old." As the forty-year-old current Commander of the 1st Engineer Battalion, I stood in awe, and still do, of Colonel Gara and the Greatest Generation.

History proved to be a powerful motivator.

WHAT I LEARNED

1. **Sports can be a powerful team-building tool**. Sports can be a very powerful team-forging tool and if used correctly can not only help teams achieve their goals but also can build overall confidence and boost motivation.

2. **Competition**. Competition within an organization can be a powerful motivator and driver for successful goal achievement.

3. **Certification is important in order to train and build confidence in individual teammates.** These certified leaders will believe in themselves because they know their craft and can execute under pressure.

4. **History**. History, if used judiciously, can also be a significant source of inspiration. Remembering and reliving the achievements and successes of the past can drive desire for success in the present and in the future.

Those who dare to fail miserably can achieve greatly.
~PRESIDENT JOHN F. KENNEDY

CHAPTER 4

FROM MISSION IMPOSSIBLE TO MISSION ACCOMPLISHED

*Individual commitment to a group effort—
that is what makes a team work, a company
work, a society work, a civilization work.*

~VINCE LOMBARDI

THE REMARKABLE STORY OF THE DAVIDSON-STYLE SEA HUT

This chapter tells the remarkable story of the Davidson-Style SEA Hut, which started as a modest shelter for troops, but turned into a powerful team building and a life-saving tool that in turn launched a successful career and life's work. How this came about was due to the vision of two very different, but equally innovative and persistent men—one a senior, war-tested Major General, the other a young, inexperienced 2nd Lieutenant. Together these two men, each in their own way, made a valuable and significant contribution to U.S. military success.

It all started with what seemed to be a "mission impossible" idea.

There may be times in your career when you find yourself working for someone who others will often say is a "dreamer" because their ideas are far beyond conventional thinking, seemingly impossible to ever achieve. If that's the case, count yourself very lucky; these are the leaders who use every single resource available to them, no matter how unconventional, and who are open to innovative thinking for the sake of their team. The "dreamer" bosses who envision the impossible missions are very often gifted leaders whose ideas lead to outcomes that foster teamwork, perserverance, and ultimately, successful solutions.

In 1998 I found myself with just such a "dreamer" leader.

The setting was Bosnia from 1998 to 1999. The 1st Armored Division, led by then Major General Larry Ellis, deployed much of the division to Bosnia for a six-month tour of duty. Major General Larry Ellis would later go on to become 4-Star General Ellis. Bosnia and Herzegovina was a war-torn country. It seemed like every square foot was laden with hazardous minefields. Every village and town was filled with families torn apart. The economy was broken. The government, whose efforts at rebuilding were plagued by distrust among the disparate communities of Bosniaks, Croats, and Serbs, was essentially dysfunctional.

And then, during the Christmas holidays, President William J. Clinton visited our headquarters in Tuzla. He visited to thank the troops for their efforts, but also to let us know that we were not coming home as originally scheduled. We would be staying for another six months. It would be a yearlong deployment. That meant another six months of living in tents. Once we left Bosnia, the 1st Cavalry Division would replace us and live in tents. The big unknown at that point was not knowing how long U.S. forces would be in Bosnia, and that's when Major General Larry Ellis' "impossible" idea was born.

THE "IMPOSSIBLE" MISSION

Major General Ellis decided that if troops were to be in Bosnia for an unknown period, they needed better living conditions. They needed indoor billeting, cafeterias, a hospital, a new runway, a new port of entry into Europe, and much more. But it wasn't enough for General Ellis to just build better accommodations for the troops, he wanted more—he wanted accommodations that would not only provide better living conditions but would also be designed in such a way as to unify, strengthen, and solidify the all-important concept of building teams. Major General Ellis was a leader who was committed to team building. He understood the importance of a team bonded together through trust and shared experiences.

This "mission impossible" was turned over to our team of engineers.

As the Commander of the 1st Armored Division Engineer Brigade, the construction project was my responsibility. I had the resources of our own troops to fulfill this mission, as well as the resources and support of our many allied countries and our civilian contractors.

One of the first goals was to build facilities so that the troops would be living indoors and no longer in tents. Major General Ellis had a clear vision for what he wanted. He said that he wanted something like the Navy Southeast Asia huts, known as SEA huts. But he also wanted to maintain the all-important unit integrity. A typical SEA hut could perhaps fit a squad of eight to ten soldiers. However, General Ellis wanted more. He wanted to maintain squad, platoon, and company integrity. There are three squads of eight to ten soldiers in a platoon, and three to four platoons in a company. And his vision didn't stop there. He also wanted the troops to have overhead cover, so they were somewhat protected from the rain and snow when walking to the latrine at night. General Ellis was a demanding leader, but among his many great qualities, he cared for his troops and wanted to build great teams, thus the importance of the design elements that he outlined.

The SEA hut has been around since the Vietnam War and provided the troops with shelter, in peace and in war, without any issues. The fundamental design was sound, functional, and time-tested. There was really no need to change the design, except that General Ellis wanted living spaces that also connected teams together. He truly understood the value of unit integrity and how living with each other might help to build the bonds that could later save lives. The final design had to achieve both purposes: comfortable shelter and unit integrity. My group of professional engineers that included both military engineers and civilian contractors went to work on the design. We tried multiple versions of unit-designed SEA huts, but General Ellis was not satisfied with anything we presented, and he sent us back to the drawing board again and again.

And then, one evening after another unsuccessful effort of trying to convince General Ellis of the new SEA hut design, everything changed.

I had scheduled a meeting with our team to discuss our options for this increasingly challenging mission. It was at that meeting that mission impossible began to change into mission possible. The unlikely change agent was a young officer, 2nd Lieutenant Ross Davidson, who was sitting far in the back of the room. He raised his hand. Stood up. And then 2nd Lieutenant Davidson told the rest of the senior engineers, the civilian contractors, and me that he thought he knew what General Ellis was looking for, and that he could design it.

I first asked him to tell me about his background. He said that he had just graduated from Notre Dame with a degree in Architecture. I challenged him saying, "Okay, 2nd Lieutenant Davidson, what you're saying is that I've been using the great ideas of all these senior military professional engineers, as well as our civilian engineers, who together cannot seem to come up with the answer, but you, a 2nd Lieutenant with no experience in the Army, think that you know what the General wants?" Without any hesitation, 2nd Lieutenant Ross Davidson said, "Yes Sir, I do."

CONFIDENCE AND INGENUITY FROM THE LEAST LIKELY SOURCE

There was something about 2nd Lieutenant Davidson. It was more than just his self-assured confidence; it was his polite, quiet, unwavering certainty. I asked 2nd Lieutenant Davidson to lay out the design he had in mind that would satisfy General Larry Ellis, and we would talk. Then, I went back to work with my senior engineers to try to further refine the designs they had been working on.

Several days later there was a knock on my door, and there stood 2nd Lieutenant Davidson with his drawings. I welcomed him in to review his designs. I was very impressed. He had designed the SEA huts in such a way as to achieve everything that General Ellis had requested. I told 2nd Lieutenant Davidson that he was going to the next meeting with me to brief General Ellis. At that meeting, Ross Davidson laid out the blueprints of his design and walked General Ellis through the blueprints step by step. At the end of the briefing, General Ellis looked at me, and announced, "That's exactly what I'm looking for. Let's start construction."

We went to work building the unit-centric SEA huts using the first new design revision since they were originally introduced during the Vietnam War. When the top Admiral from the Navy visited Bosnia, he looked at the new SEA hut design and was so impressed with not only their functional space but their team-building design he announced that this SEA hut design would be the new standard. To honor 2nd Lieutenant Davison's effort, I started to call this new version of the SEA hut, the Davidson-Style SEA hut. I did not know if it would stick, but it did. And from then on in the military construction manuals, the Davidson-Style SEA hut became the standard.

ANOTHER SURPRISE

But that wasn't the end of 2nd Lieutenant Ross Davidson and his story. Our paths would cross not once, but three more times. Once in person while still in Bosnia, and once by reputation, and again in person shortly before my retirement from the Army.

The first encounter came shortly after the new SEA hut design. 2nd Lieutenant Davidson came to my door again. He wanted to thank me for giving him a chance to provide his idea even though he was the most junior and inexperienced person in the room. I told him how proud I was of him and thanked him for speaking up. I then suggested he would make a great senior engineer one day if he stayed in the military. But he surprised me again with that same quiet confidence and certainty. "Sir, I do not want to be an Engineer Officer. I want to transfer from the engineer branch to the Medical

Service Corps branch." I asked him why in the world he would do that? Ross, as expected, had done his homework. He believed his architectural talents could be put to good use in the Medical Service Corps. While I would have loved for him to remain an Engineer Officer, I shared with him a principle in which I believe deeply—that everyone must choose their own path in life. And I added that I would certainly support his decision and wished him well.

More than two decades later came my next encounter. It was with 2nd Lieutenant Ross Davidson by reputation. At the time, I was the Commander of the U.S. Army Corps of Engineers, and we were building many new hospitals for the Department of Defense. During that time, I visited the headquarters of the Surgeon General of the Army.

All along the corridor walls of his headquarters were photos of completed hospitals and artist renditions of those being planned. As I walked the halls with the Army Surgeon General, looking at those architectural pictures and renderings, I suddenly thought of 2nd Lieutenant Ross Davidson, the "mission impossible" military architect.

I asked the Surgeon General if by chance he had ever heard of a military architect named Davidson. The General stopped in surprise and looked at me, "Do you mean Lieutenant Colonel Ross Davidson?" My response was a big smile and the words, "Yes, Ross Davidson. Do you know him?" The Surgeon General looked at me and said, "Tom, most of the hospitals on these walls were designed by Ross Davidson."

I stood there and thought about that quiet, confident, unassuming young man—a 2nd Lieutenant no longer—who possessed the self-assurance to stand up and say, among all those more senior engineers, "I have the answer." I thought of our last meeting together when that same young officer made the call about his path in life for himself. I thought about all the troops and family members who benefited from the hospitals he designed. And then I thought back to General Larry Ellis who had started the ball rolling by having a vision that not only built housing and strong teams but inspired ideas as well.

Shortly before my retirement from the Army, I decided that I wanted to see Lieutenant Colonel Ross Davidson again. Ross came to my office, and we reminisced about our days in Bosnia together. I told Ross that when he stood up with his idea on a new SEA hut design, he taught a much older officer, Colonel Tom Bostick, a powerful lesson in leadership. Even the youngest and most inexperienced among us can make significant and lasting contributions to team success.

WHAT I LEARNED

1. **Leaders with vision.** The first lesson is one that was well expressed by American inventor, engineer, and business-man Charles F. Kettering: "A problem well-stated is a problem half-solved." Leaders with vision who can state that vision clearly, like Major General Larry Ellis who

wanted living areas that would go beyond simple shelter but would serve as a life-saving team-building tool, stated his vision clearly.

2. **Look for fresh ideas from the young and inexperienced**. The second lesson is that the young and inexperienced have a lot to offer. They often provide fresh ideas, turn existing models inside out and upside down, and have the energy to execute those ideas and turn them into workable solutions. Leaders must create an environment that allows all members of the team to feel comfortable in speaking up—especially those least experienced. The secret sauce that the best leaders add to their mix is trust. In a climate of trust, even the most junior team member feels that his or her contribution will be valued and considered with respect.

3. **Respect individual passion.** The third lesson learned is that while a team is a powerful unit, it is made up of individuals who must make their own decisions regarding their path in life. And if an individual has a passion and does the necessary homework then they will succeed on any team—just as the young, untried, but talented 2nd Lieutenant Ross Davidson made an impossible mission possible.

It always seems impossible until it's done.
~NELSON MANDELA

CHAPTER 5

THE PORT OF RIJEKA, TUZLA AIRFIELD, AND KOSOVO

The most admired leaders are able to build and manage teams that can overcome the obstacles faced in creative projects.

~SCOTT BELSKY

We all face challenges every day. Some of these challenges are significant. There's the challenge of relocating you and your family to a new town or starting from scratch in a new job. Now consider the challenge of ramping up an entire port capable of supporting massive military demands. Or consider the challenge of lengthening a small airport to a strategic airfield capable of international travel in just ninety days. Those were the two massive challenges we faced, and this is the story of how through inspired leadership, sheer hard work, finding the best talent, and tight teamwork, we successfully overcame both challenges.

THE PORT OF RIJEKA

The first challenge was presented not long before the 1st Armored Division was preparing to depart Bosnia following a yearlong deployment. One afternoon, our Division Commander, Major General Larry Ellis called our Assistant Division Commander, our logistician, and me into his office. He had a question. He wanted to know why the military always used the European ports of Rotterdam in the Netherlands and Bremerhaven in Germany to bring in equipment for troops that were serving in Bosnia, especially since Rotterdam and Bremerhaven were each over one thousand miles from Bosnia. We pointed out that this was the way equipment had been brought into Europe for fifty years. And that for fifty years these two ports were successful. They just worked.

General Ellis then threw another question at us. He asked whether it was more expensive than a closer port to bring equipment into Rotterdam and Bremerhaven and then load everything onto trains and ground transportation to deliver all this equipment to Tuzla, Bosnia, where it was needed. We confirmed that using those distant ports was expensive. And then he asked, "But what was the alternative?" That turned out to be the key question that would initiate a seemingly impossible mission—move the port of debarkation for the 1st Cavalry Division to a closer port. We would change the port from which the 1st Cavalry Division would exit their vehicles and equipment for onward movement to Tuzla, Bosnia.

General Ellis said that there was an alternative to the ports in the Netherlands and in Germany. He pointed out that there was a port just 215 miles away, almost a thousand miles closer. It was the Croatian Port of Rijeka.

This was a bold move from a visionary leader.

The first step was to do some reconnaissance. We had to see what we were dealing with, the possibilities as well as the impossibilities, so we jumped on a Blackhawk helicopter and flew to the Port of Rijeka. The flight seemed to take forever. After we landed, my assignment was to look not only at the port itself but at the railroad tracks and the infrastructure that served it, and to report my findings to General Ellis. Our logistician was tasked with reviewing the warehouse facilities adjacent to the port. We went to work.

After a few hours, General Ellis asked for our thoughts. I expressed concern that the port looked almost abandoned. There were very few workers on site. One issue was whether the port had the staffing required. And while I agreed that one of the great advantages of the Adriatic Sea Port of Rijeka was that it was much closer to Tuzla, Bosnia, almost a thousand miles closer than the Dutch and German ports, I also expressed concern that significant issues could arise out of the logistical challenges to make this port and its infrastructure capable of supporting the 1st Cavalry Division. Our logistician made similar logistical observations and expressed his concerns. We returned to the Blackhawk for another long flight back to Bosnia.

After landing in Bosnia we thought that was the end of it, that the port of Rijeka option would be dropped. We couldn't have been more wrong. Not only did we get an unexpected but powerful lesson in leadership, but we were thrown right into one of the largest and most innovative changes in military logistic procedures in fifty years.

General Ellis told us to join him in a Video Teleconference Conference (VTC) with the NATO Stabilization Force Commander, who at the time, was based in Sarajevo. General Ellis began the conference by saying that he had just returned from the Port of Rijeka, and his logistician and his engineer (that was me) had convinced him that it was possible to bring the 1st Cavalry Division into the Port of Rijeka rather than Bremerhaven or Rotterdam. This port change would save considerable money and time, train the Army to use unfamiliar ports, and would also help rebuild the economy of Croatia. I couldn't believe what I had just heard. And glancing over at our logistician, it seemed that neither could he. But the NATO Stabilization Force Commander was delighted and approved the recommendation to move to the new port right away.

After the VTC, I asked General Ellis how I had convinced him that it was possible to use the Port of Rijeka. I remember apprising him of the possibilities but also of the many, yet to be determined, challenges. I never remembered saying that it was possible. But, General Ellis responded with, "True, but you did not say it was *impossible*." I started to list some of those challenges in the hopes that the General would reconsider his decision. I told him we hadn't checked the railroad tracks to see if our equipment could be transported to our base in Bosnia. I pointed out that the tunnels might be too low and too narrow to accommodate the 1st Cavalry Division's tanks and other equipment. I explained that we hadn't even checked whether the bridges could handle the capacity of the tanks. General Ellis's response? He simply said, "Okay, then you should go and check the railroad tracks, tunnels, and the bridges."

So, we did. We inspected the entire route of travel from the Port of Rijeka to Tuzla, Bosnia. We inspected the railroad tracks, the tunnels, and the bridges to see if the Port of Rijeka could serve as the alternative that would change half a century of military practice.

We returned, but sadly not with the good news General Ellis was expecting to hear from his experts. On our return, I had to tell General Ellis that "his" plan to use the Port of Rijeka would not work. It was impossible.

True to form General Ellis asked why. I was ready with my list of challenges: first, the bridges could not handle the weight of the M1A2 Abrams Tanks we needed to transport; second, the tunnels were too small to accommodate our M1A2 Abrams Tanks and M2 Bradley Infantry Vehicles. I thought that these two challenges alone would end the discussion of using the Port of Rijeka. But once again, I had underestimated General Ellis's vision and determination. And once again,

General Ellis provided an innovative solution to the challenges that had just been outlined for him. His solution? He decided that only the equipment that could be accommodated by the country's trains, tunnels, and bridges would be shipped from Fort Hood, Texas, to the Port of Rijeka. The heavier equipment like the 1st Cavalry Division's M1A2 Abrams Tanks and M2 Bradley Infantry Fighting Vehicles would remain behind in Fort Hood, Texas. We would leave our tanks and Bradleys in Bosnia to be used by the 1st Cavalry Division.

This was the first time that the Army would ship their wheeled vehicle and some track vehicle equipment into an immature port. It was also the first time that units would leave equipment behind, significant amounts of their equipment that the incoming units would use.

General Ellis's idea of deploying into a port like Rijeka and leaving equipment behind would serve the Army well in the years to come, and his actions would illustrate his skill as a visionary leader. His idea would echo the advice once offered by Robert A. Heinlein, an American writer, aeronautical engineer, and retired Naval officer who said, "Always listen to experts. They'll tell you what can't be done, and why. Then do it."

TUZLA AIRFIELD

General Ellis was a leader who questioned everything for the benefit of both his troops and the mission. He always managed the government's money as if it were his own. He would have approved of Winston Churchill's advice: "To improve is to change; to be perfect is to change often." And so it was that he challenged us with another seemingly impossible mission that pushed us to the limits of our ability and stamina but showed us what amazing feats we were capable of if we worked as a team.

It was not long after the final decision to use the Port of Rijeka as the port of disembarkation for the 1st Cavalry Division that General Ellis and I met once again—this time to discuss the Tuzla, Bosnia Airfield. As always, General Ellis started with a question. He wanted to know why the 1st Cavalry Division was flying into airports in Frankfurt, Germany, or Sarajevo, Bosnia and Herzegovina, and not into Tuzla where we were based which was much closer. I recall explaining to General Ellis that Frankfurt and Sarajevo were both strategic airfields and Tuzla was not.

Then he fired off another question. He asked what it would take to make the Tuzla airfield a strategic airfield so that the 1st Cavalry Division could fly directly from Fort Hood, Texas, to Tuzla, Bosnia. Convinced that he didn't yet grasp the scope of what he was asking for, I fell back on a clear and hopefully convincing analogy of why a strategic airfield at Tuzla was not possible. "General Ellis," I said, "you've served in Washington, D.C. You know that Washington, D.C. has two major airports, Dulles and Reagan. When the large 747 planes fly into Washington, D.C., they must land on a strategic airfield with approximately 10,500 feet of runway to accommodate them. Dulles is a strategic airfield, and that's why all 747s fly into Dulles. Reagan is not a strategic airfield. It is a shorter airfield, so the 747 planes cannot and do not land at that airport. Here, at Tuzla, we have the equivalent of Reagan Airport, not Dulles. The 1ˢᵗ Cavalry Division cannot fly from Ft. Hood with its massive transports and equipment and land at the Tuzla airfield."

Unfazed, General Ellis looked at me, and without any hesitation, said, "I want Dulles. I want Dulles here at Tuzla."

I couldn't believe what I had heard. I rolled out an impressive array of reasons why he could not have Dulles at Tuzla and would have to settle for Reagan. The 747 planes would have to fly from the United States and land in Frankfurt or Sarajevo. To reinforce my argument, I pulled out a map of the airfield and pointed to the end of the runway. The map clearly showed a solid line of trees at the end of the runway, and beyond the trees, homes, and beyond the homes, a railroad track. The first challenge was that all these items would have to be removed. And then once the removal was complete, we would face the challenge of expanding the smaller airport at Tuzla into a massive strategic runway that would support the 1ˢᵗ Cavalry Division's troop and inbound air transport and inbound equipment. And finally, the third and most difficult challenge was the challenge of time available prior to the deployment of the 1ˢᵗ Cavalry Division. I explained that it was currently August, and we were leaving Bosnia in October, in just a little over two months. We wouldn't finish the construction before the 1ˢᵗ Cavalry Division had to be in

Tuzla. "General Ellis, it's just not possible," I asserted, confident that the arguments I had so skillfully marshaled would convince him to abandon his idea.

But I was wrong. General Ellis put his hand on my shoulder and told me that he had great confidence in me. Then he asked me to tell him exactly what I needed to build him "Dulles" airfield at Tuzla.

Okay. I had my mission and I never, ever said no to a mission once a decision was made. General Ellis made his decision. I went back to my office, called the head engineer in Germany, and told him our plans to construct a strategic airfield in Tuzla. He said this was crazy. He said I had to convince General Ellis that this was a bad idea and that we were certain to fail. I assured him that I tried, but I couldn't budge General Ellis a single centimeter from his idea. I told the senior engineer in Germany that we were going to figure out a way to execute this mission successfully.

First, I started to search for an expert in airfield construction. We had some U.S. Air Force leaders in Tuzla, and I asked them if they knew of anyone with experience in airfield construction. They told me about an Airman in Ramstein, Germany. The next day, I flew to Ramstein Airbase and found him, Master Sergeant Patrick Daize.

"I understand that you know a lot about constructing airfields," I said.

Master Sergeant Patrick Daize replied, "Yes, Sir, I've built several of them."

I said, "OK, I need you to go with me."

He looked puzzled and asked, "Go with you where, exactly?"

"To Bosnia," I told him. "I need help building 'Dulles' airfield at Tuzla."

Without hesitating he announced, "I'm in."

After a quick check with his boss, he and I were in a plane and off to Tuzla. During the flight, I asked what was the worst thing that could go wrong with the airfield construction. Master Sergeant Daize explained that quite a lot could go wrong. There were a lot of moving parts, and any one of them could undo the project. The first thing we need to make sure of, he pointed out, was that we prepare the asphalt mix correctly, and to do that, we needed an asphalt mix expert. I called the U.S. Army Corps of Engineers and asked that they send me the best person they had in asphalt mix design and gravel quarry work. Before I knew it, our asphalt expert quickly deployed into Tuzla.

The next two months were grueling. We worked day and night, seven days a week, 24–7, non-stop.

Then, in early October, I sat in a chair on the tarmac next to General Ellis, the incoming 1st Cavalry Division Commander, and the NATO Stabilization Force Commander as the first 747 aircraft with troops flying directly from Fort Hood, Texas, landed at Tuzla, Bosnia. Our "Dulles" had been built!

General Ellis had a vision of what was possible. And, while to many of us, his visions seemed out of reach—impossible—he showed us that we could achieve the impossible. It takes leaders like General Ellis to provide the leadership, inspiration, and resources to those who must execute and deliver on these seemingly impossible missions. When there are challenges, and it looks like all is lost—and we had plenty of those—there are still ways to win.

The notion of doing something no one has ever done before is not something that most leaders around General Ellis would have thought of or recommended. The life of a leader can be a lonely place until one builds a team of believers. Although it took some time, General Ellis made me a believer. I believed in him. And he believed in our team and me. I was able to build a team of my own which in turn exceeded all expectations.

Years later as a senior General Officer, I'm sure some of those around me thought I was a "dreamer" with my own set of impossible missions. I learned early in my career that not much is impossible if you have the right leadership, team, and resources in place.

For a leader, it is important to keep pressing beyond any hurdles or what looks like insurmountable challenges. It is important to build the right team and to ensure true experts are on the team. It is important to believe in the team and to provide the necessary resources for success. This is what turns an impossible mission into a mission accomplished. But there is more.

DIVISION WARFIGHTER AND KOSOVO DEPLOYMENT
Just prior to the 1st Armored Division's redeployment from Bosnia back to Germany, General Ellis had a meeting with

all of his seven Brigade Commanders, which included me. Four of these Brigade Commanders would go on to become General Officers. General Ellis asked us for our thoughts about training for a Division Warfighter Exercise in March of the next year. A Division Warfighter Exercise required significant preparation. An outside team of experts would evaluate the division's preparedness for war. To a person, the Brigade Commanders thought this was not a good idea. We pointed out that in order to train for a Warfighter, we would need to be back in the field in November. It was October, and we were still in Bosnia. On top of this, we made the decision to leave a significant amount of our equipment behind in Bosnia, which would make training incredibly challenging.

After all the discussion, General Ellis made his decision. We would conduct a Warfighter Exercise in March of 1999. He went on to say that our logistician and engineer (me) would obtain the necessary equipment from storage facilities in Italy. Until we had the necessary equipment, we would use smaller vehicles for training.

We conducted a very successful Warfighter Exercise in March. Then, in April, the 1st Armored Division received orders to become part of Operation Allied Force in Albania and Operation Joint Guardian in Kosovo. We were prepared. Once again, General Ellis made the right call in using the Warfighter to prepare the division for war. In both the port of Rijeka and the strategic airfield in Tuzla, Bosnia, we faced what seemed to be mission impossible. The Division Warfighter training exercise and evaluation seemed to be another impossible goal to attempt immediately following a year in Bosnia. Once again, it was the right decision. As Nelson Mandela wrote, "It always seems impossible until it's done." General Ellis brought those inspiring words to life for us.

WHAT I LEARNED

1. **Raise the bar.** I learned the power of setting high expectations. Leaders can and should challenge the status quo and the boundaries set by the past when forging a new future.

2. **Ask the right questions.** Asking the right questions can make all the difference between success and failure. Getting to the very heart of the challenge, deciding what's needed, and then providing the appropriate resources can result in success.

3. **Visionary leadership.** Leaders who have vision are critical to success, especially in disruptive times. These leaders not only envision bold solutions but envision their teams successfully executing their vision. Visionary leadership leads not only to successful missions but to motivated and confident teams.

4. **Experts.** Expertise is invaluable when tackling challenging missions on a very short timeline. Seek out and find the experts then add them to your team.

> *I am always doing that which I cannot do,*
> *in order that I may learn how to do it.*
> ~PABLO PICASSO

CHAPTER 6

ON THE POWER OF CALM AND QUIET LEADERSHIP

Leaders don't create more followers,
they create more leaders.

~TOM PETERS

A POWERFUL LESSON IN PEOPLE-CENTERED LEADERSHIP

During my thirty-eight-year career with the Army, many leaders stand out in my mind. There were different leaders who trained, guided, supported, encouraged, and molded me. There were leaders who showed me how to create and lead teams of my own. How to get the very best from the thousands of soldiers and civilians under my command. How to welcome every challenge.

How to solve problems when there seemed to be no solutions. There were leaders who mentored. Leaders who inspired. Of all the leaders there is one I have known and admired the longest.

This leader was not only a mentor from my earliest days as a newly minted West Point cadet but also an ever-present role model as my own career grew in scope and developed in responsibility and rank. He was a calm, quiet, caring, and confident leader, who always lived the concept of Mission First, People Always. Most leaders will always place the mission first, but thinking about people while in the thick of a mission is challenging for most—but not for this leader.

This leader's style of leadership is exactly what the Harvard Business Review calls "servant leadership" and points out that, "servant leaders view their key role as serving employees as they explore and grow, providing tangible and emotional support as they do so." This is echoed in an article published in *Inc. Magazine*: "a leader who puts others first creates an uplifting, motivating culture that inspires confidence."

One of the best descriptions of this leader and his calm and quiet power is best voiced by the ancient Chinese philosopher Lao Tzu, the founder of Taoism, who is perhaps best known for the quote: "A journey of a thousand miles must begin with a single step." It is this type of leadership that Lao Tzu described that has lasted through centuries:

The superior leader gets things done with very little motion. He imparts instruction not through many words but through a few deeds. He keeps informed about everything but interferes hardly at all. He is a catalyst, and though things would not get done well if he weren't there, when they succeed he takes no credit. And because he takes no credit, credit never leaves him.

SERVICE AND SACRIFICE

This leader and his wife have known me longer than anyone else during my time in the Army. As a West Point cadet, I often went to their home on weekends. And throughout my career, even after I was married, this leader and his wife kept in touch with me and my wife Renee. But during all those years I had never actually worked for him, or experienced his unique style of leadership until he became the Chief of Staff of the Army, and assumed the position in which he was often referred to as simply "Chief." During this period, my role was that of Executive Officer to the Chief of Staff of the Army. And it was in that position that I was able to observe his many qualities of leadership. One quality that I saw time and time again was the care that the Chief had for people. The Chief's emotional intelligence was the best that I had ever witnessed. He gave credit generously. He never wanted credit for anything. He and his wife were the ideal examples of service and sacrifice to so many in our Army and Nation. When I think about leadership, I think about this Army Chief of Staff. Above all, this Army Chief of Staff is a soldier who has great vision, the skills to operationalize that vision, and the qualities and attributes to motivate and inspire both soldiers and civilians to accomplish what might be seemingly impossible.

THE TRUE STORY OF ONE ENSIGN DAN JOHNSON

Early one morning, on his way to The Pentagon, the Chief called me from his car.

"Tom, good morning. Have you read the story by George Will about Ensign Dan Johnson?"

I replied, "Sir, I read the article. Ensign Johnson is quite a remarkable young man."

The Chief said, "I want to visit Ensign Johnson today."

I reminded the Chief that he was scheduled to testify before the House Armed Services Committee about the upcoming Army budget. The budget was a critically important item and of the highest possible priority. There could be no distractions. But the Chief was not to be moved.

I tried again, this time suggesting a compromise. "Sir, perhaps you can visit with the ensign later today after you finish testifying about the Army budget before the House Armed Services Committee."

The General was still not to be dissuaded. "No, I'll go before the Hearing." And the Chief asked me to check to see if another four-star General would like to join him.

The two Generals have similar leadership qualities. Both are calm, caring, and compassionate, yet they are both decisive, courageous, and battle-hardened warriors. They both share one other attribute which I will reveal later.

I called the retired four-star General. "Good morning, Sir. The Chief would like to visit Ensign Dan Johnson at Walter Reed this morning, and he would like you to go with him if possible."

The retired General agreed, "Tom, sure, I'm happy to join the Chief."

I tried again to change the schedule and move the meeting with the Ensign to later in the day, after the Congressional obligation. "Sir, I was hoping that you might mention to the Chief that he could go to Walter Reed after he testifies before Congress today," I suggested.

The retired General paused, "What does the Chief want to do?"

I said, "The Chief wants to see Ensign Johnson this morning before the hearing."

The retired General replied, "Tom, then that's what we'll do."

As we prepared to dispatch the car that would take both generals to their meeting with Ensign Dan Johnson, I couldn't help but replay in my head the Ensign Dan Johnson story that I had just read. Because, as it turned out, that story cast a very bright light not just on one man, not just on two, but on all three.

WHO WAS ENSIGN DANIEL H. JOHNSON?

Ensign Daniel H. Johnson had recently graduated from the University of North Carolina at Chapel Hill in the summer of 1999 and received a commission as a Naval Officer. Dan was serving aboard the U.S.S. Blue Ridge designed originally as one of two amphibious command ships and commissioned as the flagship of the Seventh Fleet.

On the morning of August 23, 1999, Ensign Johnson was the Safety Officer on board the U.S.S. Blue Ridge when Korean tugs guided the ship into position to leave the port of Pusan in South Korea.

And that's when everything in Ensign Johnson's world changed.

One of the ropes from the tugboat began shaking violently and the seamen scattered from the immediate area of danger. However, a line entangled one of the legs of Seaman Steven Wright and started to drag him toward the chock. Chocks are structures on a ship that are designed to guide the mooring ropes smoothly from the ship to the shore or to other ships. They are built and reinforced to withstand any structural damage to a ship while it is moored. If Seaman Steven Wright was pulled against the chock, the impact would most likely prove fatal.

That's when Ensign Dan Johnson sprang into action. He reached Seaman Wright to try to free him from the deadly pull of the rope, but not before the rope had severed Seaman Wright's foot and four fingers.

What drove Ensign Dan Johnson to race toward his endangered shipmate? When asked later why he did it, Ensign Johnson simply said, "I just ran over there. It was just kind of instinct."

But that wasn't the end of the story.

Ensign Johnson found himself tangled in the same rope and within seconds, that line had severed both his legs below the knees and his left pinky finger.

LIVING BY THE MOTTO "MISSION FIRST, PEOPLE ALWAYS"

Ensign Johnson was initially treated in the hospital in South Korea but had arrived at Walter Reed, and the Chief wanted to be there to welcome him home on the same day as the Chief's scheduled Congressional testimony.

Why was this one act so important? Why was it more important even than one last meeting to prepare for a Congressional Hearing? What was so special about this injured ensign? Why did the Chief invite a retired four-star General to join him in meeting with Ensign Dan Johnson?

Part of the reason was that both Generals are amputees from their service in Vietnam.

But that wasn't the only reason these two leaders went to meet a junior officer in another service, the Navy. The Chief wanted to meet with Ensign Johnson and his parents and let them know that not only would Walter Reed Army Hospital provide excellent care, but to reassure that brave officer that even as an amputee, he could have a very successful career in the military and in life. The two Generals wanted to inspire Ensign Johnson at this incredibly challenging time in his young life.

LEADERS DON'T LEAVE UNTIL THEIR WORK IS DONE

I had hoped that the Chief would spend only an hour at Walter Reed and then return to The Pentagon so the Army Staff could conduct one last preparatory meeting for the Congressional budget testimony. The Chief had already attended several preparation meetings, or "murder boards" as we called them. These were intense preparatory sessions where we asked the Chief every possible question that we could imagine he might face during his testimony. The Chief studied meticulously for hearings. His standard was that there should never be a question raised by a member of Congress that we had not anticipated and prepared for, and there rarely was. That morning, we had one additional preparation meeting scheduled.

We waited. We waited for the Chief to return to The Pentagon that morning, but he did not return. He did not attend the last "murder board" before his Congressional testimony later that day. Both Generals spent that morning at Walter Reed with Ensign Johnson, his parents Wallace and Sallie Johnson, and with Seaman Wright who was in the room down the hallway.

Finally, after several hours, the Chief called me at The Pentagon.

"Tom, I suppose that I do not have time to return to The Pentagon for that last preparation meeting you scheduled."

"No, Sir. Your driver will take you straight to Capitol Hill for your hearing," I replied.

The Chief said, "That's okay," and with much excitement in his voice, he continued, "Ensign Johnson prepared me. What an inspiring young man. His father is pastor of First Presbyterian Church, and he was also an Army chaplain who spent twenty-seven months in captivity as a prisoner of war and earned the Distinguished Service Cross and the Silver Star. His mother, Sallie, is a teacher. What a wonderful family!" I recall the Chief telling me that he asked Ensign Johnson what made him react in the way that he did to assist his fellow sailor. Ensign Johnson responded, "Isn't that what leaders do?"

"You know," the Chief continued, "I had gone to Walter Reed to reassure Ensign Dan Johnson that the doctors and medical professionals would take care of him. We wanted to tell him that he could do anything he wanted to do in life, even as an amputee, but as it turned out, Ensign Johnson was the one who inspired us."

Ensign Johnson received the Navy and Marine Corps Medal for heroism while serving as Safety Officer on Board U.S.S. Blue Ridge at Pusan, Korea, on August 23, 1999.

Later, in a telephone interview during his recovery, Ensign Johnson said, "I'm pretty confident I'll bounce back. I'll be back on my new feet." He joked, "I don't think the Olympics are in my future . . . I'll have to use my mind rather than my body."

I'm sure it was this type of positive attitude that the two, four-star Generals witnessed when they visited Ensign Johnson at Walter Reed.

And what about the testimony before the Congressional Committee? The Chief hit a "home run" during his testimony following his visit with Ensign Dan Johnson, and the Army budget was passed by the House Armed Services Committee.

A CALM, QUIET, CARING, AND CONFIDENT LEADER

Early in the afternoon that same day, the Chief returned to The Pentagon. Except for the retired four-star General, a few doctors and nurses at Walter Reed, and the Johnson family, very few people knew that the Chief had visited Ensign Johnson and Seaman Wright that morning. Not the Army Staff, not the Navy, not anyone else in The Pentagon. The Chief is the epitome of the superior leader who "gets things done with very little motion." He has never been one to take credit and thus, the "credit never leaves him." Although the Chief is a battle-tested combat leader always focused on the mission, he understands the importance of showing care and compassion for people. He and his especially caring and wonderful wife are humble, servant leaders who led the Army with grace and in an exemplary manner.

What about Ensign Dan Johnson? Ensign Dan Johnson eventually left the military after the initial recovery from his injuries. He became an attorney and served as the Assistant District Attorney for Wake County, North Carolina, before opening his own legal practice. Ensign Daniel Johnson demonstrated true heroism and the power of resilience as he bounced back stronger from his injuries.

WHAT I LEARNED

1. **Leadership.** Leadership exists at all levels of experience and can be found in the youngest among us.

2. **Role model.** Being a role model is not linked to seniority and rank. Senior and experienced leaders can clearly serve as great role models who inspire the young; however, young people such as Ensign Johnson are equally inspiring role models.

3. **Resilience.** Ensign Johnson displayed the power of resilience and how someone can experience a tragic life changing event, recover, and bounce back as a strong and successful person.

4. **Modesty.** When you live a life of not taking credit, credit never leaves you.

> *Courage is the most important of all the*
> *virtues because without courage, you can't*
> *practice any other virtue consistently.*
> ~MAYA ANGELOU

CHAPTER 7

FIRST TEAM

Remember, upon the conduct of
each depends the fate of all.
~ALEXANDER THE GREAT

A team can be described as the delicate balance between the individual and the group. Both must be in perfect alignment. The glue between individuals, groups, and multiple groups on the same team is alignment. Alignment is achieved when the various parts of an organization have a common understanding of the organizational goals. But what happens when that alignment is tested? What happens when carefully crafted plans change? Or when exhilarating but relentless competition that, despite best efforts, ends in disappointment? Or when risks fail to yield the expected rewards? How do many individuals or teams embrace these challenges and use them to become stronger?

In this chapter, stories of challenge and the teams and individuals involved demonstrate what deep reserves of loyalty, persistence, and alignment are needed to reach the finish line.

"FIRST TEAM, SIR!"—A TALE OF TWO TEAMS

The 1st Cavalry Division is one of the premier and most modernized tank divisions in the Army. It has a long and storied history and has served the Nation well in peace and war. The 1st Cavalry Division was activated in 1921, starting as pure cavalry, patrolling the border between the United States and Mexico on horseback before transitioning from horses to tanks and armored vehicles. It is one of the most decorated divisions in the Army with forty-three of its members receiving the Medal of Honor, and two hundred and sixty-one receiving the Distinguished Service Cross. Officers and Non-Commissioned Officers serving in the 1st Cavalry Division have become some of the most senior leaders in the Army. The 1st Cavalry Division is often first in much of what it does. They were first into Manila and first to Pyongyang. In fact, they are known as "America's First Team." To this day when a soldier approaches an officer in the 1st Cavalry Division, that soldier delivers a sharp salute and says, "First Team, Sir!"

The 4th Infantry Division was created in 1917 at the time when America was entering the First World War. Known as

the "Ivy" division because of the insignia its soldier wear—
four ivy leaves in the shape of a diamond—the division
distinguished themselves at battles in the Marne and Aisne
regions of France. During the Second World War, men from
the "Fighting Fourth" were among the first to land on Utah
Beach and among the first American troops to enter Paris.
They are recognized as a liberating unit by the United States
Holocaust Museum. To this day they live by their motto
"Steadfast and Loyal."

These two divisions were both stationed at Fort Hood,
on the same military installation. And that's where this story
really begins.

There was an ongoing rivalry between these two legendary
divisions—and at times it was fierce. For example, a road
separated the divisions' areas much like a fence separates a
field. During morning runs, a member of one division would
not want to be caught running in the other division's areas.
The rivalry was that intense.

And yet, despite the rivalry, both divisions were part of the
larger Army team, and there were great friendships between
individuals serving in both divisions. And this loyalty and
unity to that larger team—the Army—was best demonstrated
by a father and son.

The 4th Infantry Division was commanded by an officer
who would later serve as the Chief of Staff of the Army, the top
position in the Army. The Division Commander's son served
in the very prestigious 1–7 Cavalry Squadron in 1st Cavalry

Division. Major General George Armstrong Custer once led the 7th Cavalry Regiment. The father and son, although they served in different divisions, both of which were planning for a combat deployment into Iraq, shared a higher loyalty which aligned them both—loyalty to the U.S. Army.

As both divisions continued to train, without warning, carefully crafted plans and months of training were turned upside down, which deeply affected both divisions. But it was how these two competitors handled the change, the disappointment, and even the role reversal, that demonstrates the importance of alignment.

The war plan called for the 4th Infantry Division to move into Iraq from the north, through Turkey, while the 1st Cavalry Division would enter Iraq from the south.

By this time both divisions had completed training and were ready to deploy to serve the Nation. But a change in plans occurred. The 4th Infantry Division, instead of deploying from the north as originally planned, would now deploy from the south, taking the mission originally assigned to the 1st Cavalry Division. That was unsettling for some. But the bigger disappointment came with the orders for the 1st Cavalry Division. Their deployment was pushed back a year. And to make matters more challenging, instead of deploying, the 1st Cavalry Division would now assist the 4th Infantry Division with its deployment from Fort Hood to Iraq. The 1st Cavalry Division would stand by, watch, and keep training. It was a bitter disappointment to many members of the 1st Cavalry

Division. Some felt that they were not the "First Team" going into Iraq as their motto implied.

But that would all change . . . and that takes us to our second story: how this "First Team," through persistence and a higher loyalty, would show that disappointment is only the flip side of satisfaction.

TALENT VERSUS TEAM—A PRESCIENT LEADERSHIP DECISION

While the 4th Infantry Division deployed into battle, the 1st Cavalry Division remained behind at Fort Hood to train.

During my first year with the division, I was assigned as the Assistant Division Commander for Maneuver of the 1st Cavalry Division, a rare opportunity for an Engineer Officer. I focused on the war fighting capability of the division. We trained throughout the year with the most challenging training occurring at the National Training Center (NTC) where we "battled" against the vaunted Opposing Forces (OPFOR).

During my second year with the division, I was assigned as the Assistant Division Commander for Support. We also had a new Division Commander who would go on to become a four-star and the second most senior General leading the Army as the Vice Chief of Staff of the Army.

Of all the mock battles we engaged in, one stands out. It was after that battle that our Division Commander made a key decision, which at the time may have seemed unusual but that would, in the end, turn out to be the absolute right decision.

It all started during a session of post-battle After-Action-Review (AAR) after one of our brigade combat teams had lost a battle at the National Training Center. The AAR serves as an opportunity for teams to discuss what went wrong, why, and what could be done to ensure that it did not happen again in the future. While some leaders in business may not use the term AAR, they comprehend the importance of understanding why mistakes happen and the importance of learning from those experiences. As Ginny Rometty, CEO of IBM once stated, "I'm a big believer in lessons learned. Constantly with the team, we go over, Why? Why? Why? . . . the only bad mistake is a mistake you don't learn from." Ed Catmull, co-founder of Pixar, uses "post-mortems" with his team to learn from each movie. Our Division Commander, pointing to the horizon, illustrated the flaw in the strategy that had cost the brigade their victory. He told the assembled brigade leaders that if they had placed a scout on the high ground, they would have had good visibility on the enemy. That visibility would have made the difference between victory and defeat against the OPFOR. All teams training at the National Training Center want to win, not lose, so the brigade planned to make the necessary adjustments.

After the AAR and just when the team was feeling defeated, a young Battalion Commander, a Lieutenant Colonel who would later become a Lieutenant General—who was carrying over sixty pounds of full-field gear—showed great initiative by running to the top of the steep ridgeline to better understand the Division Commander's comment.

At the time, this Battalion Commander commanded 1–5 Cavalry. He was considered by many to be the best Battalion Commander in the division, and certainly the most outstanding warfighter. In fact, our Division Commander had considered making the 1–5 Cav Commander the Division G-3, responsible for all operations in the division. This was a position to which many of the best Battalion Commanders aspired. And no one would have been surprised if that honor had gone to the 1–5 Cav Commander. And then our Division Commander made what on the surface may have seemed a counterintuitive decision. He would not elevate the 1–5 Commander, one of our best leaders, to the Division G-3. He would leave him in a position where the risks to soldiers were high. The reason our Division Commander gave for this decision? "We cannot move the 1–5 Commander because the troops deserve his leadership in combat." It was absolutely the right decision.

Talent management is such an important skill for a leader, and our Division Commander's recognition of the best placement for the 1–5 Cav Commander, while perhaps not clearly understood by some at the time he made it, would prove to be prescient in the months to come. As Pearl Zhu once stated, "Every person is unique; put the right people with the right capability into the right position to solve the right problems." Our Division Commander understood that the right position for the 1–5 Cav Commander was with the troops who would be tested in the crucible of battle very soon.

A few months later the lead units deployed from Kuwait into Iraq. Under the leadership of their Battalion Commander, 1–5 Cav was fighting their first battle in Sadr City. It was during that first battle that our Division Commander's decision, that 1–5 Commander would be kept with his unit because "the troops deserved his leadership," proved to be absolutely correct. Even though, during the first battle, eight soldiers were tragically killed and fifty-two wounded, the 1–5 Cav Commander's leadership was one of the key elements of success in this battle of Sadr City. The 1–5 Cav Commander's presence close to the troops would be prominent in the many battles to come throughout the year-long deployment, and many men would be saved and kept safe under his leadership.

Sometimes the decision to keep your best people in the assignments of most risk is better than elevating them to positions that might be considered more important for the organization.

This takes us to the end of this story. It also leads us to another story—an additional story that demonstrates that even in the face of competition, the higher loyalty is in the end to the larger team—the Army.

THE UNIFYING POWER OF FRIENDSHIP—THIS STORY BEGINS WITH A FUNERAL

Back at Fort Hood, 1st Cavalry Division which had been left behind to stand by and watch the 4th Infantry Division, successfully deploy into Iraq, trained, and waited for their turn to be "first" once more.

But even though the 1st Cavalry Division was far away from the war while at Fort Hood, it did not take long for the war to come to 1st Cavalry Division. All too soon news of casualties reached the waiting spouses and families at Fort Hood. One of the early fatalities was a First Lieutenant, who was also an Infantry Officer and a Bradley Platoon Leader. The Chief of Staff of the Army had implemented a policy that a General Officer would represent the Army at the funeral for each fallen soldier. I was that General Officer who would represent the Army for the funeral of this young Infantry Officer. But I wasn't the only one. The Army would also assign a Casualty Assistance Officer to help his family. The fallen soldier's wife mentioned a specific officer that she wanted to be the Casualty Assistance Officer.

I contacted the fallen Lieutenant's division, the deployed 4th Infantry Division, and made the request. As it turned out, the requested Casualty Assistance Officer was not deployed with the 4th Infantry Division. I followed up by asking the Corps personnel officer at Fort Hood to find this Lieutenant, whom I suspected was in the rear detachment of the 4th Infantry Division. He was not. To my surprise, the personnel officer called me. He said that the reason that the requested officer could not be found in the 4th Infantry Division was that he wasn't part of the 4th Infantry Division. He was assigned to the Signal Battalion of the 1st Cavalry Division.

I could not believe this. Knowing the competitive nature at various levels of these two divisions, how was it that a Lieutenant from the 1ˢᵗ Cavalry Division would be the person requested by the young widow to serve as her fallen husband's Causality Assistance Officer?

The next day, the mystery was solved. Not only was it solved, but it reaffirmed the higher calling that was understood by these two officers and close friends. The Lieutenant reported to my office. His grief-stricken face spoke volumes. This was clearly a very difficult time for him. And then the story of the remarkable friendship of these two men was revealed. It turned out that they were from the same hometown in California. They went to the same college together. The Lieutenant told me about one of the last discussions he had with his friend from the 4ᵗʰ Infantry Division before he deployed. His fallen friend had said, "He knew that some of the members of the 1ˢᵗ Cavalry Division would be upset that he and the 4ᵗʰ Infantry Division would be going into battle first. However, he told me not to worry. He said there would be plenty of fighting left for the 1ˢᵗ Cavalry Division." The Lieutenant went on to say, "He told me to take care of myself and my soldiers." The two young Lieutenants then raised their bottles of beer and made a toast. They toasted seeing each other again when they redeployed.

It was not long after that final meeting that the 1ˢᵗ Lieutenant from the 4ᵗʰ Infantry Division was killed. And the two friends never saw each other again. But their friendship was stronger than any division rivalry. The Casualty Assistance Officer from the 1st Cavalry Division was subsequently able to offer his help, support, and comfort to the 4th Infantry Division's fallen Lieutenant's spouse, family, and friends.

These two officers demonstrated the true meaning of teamwork, friendship, and maturity at such a young age.

WHAT I LEARNED

1. **The organizational higher loyalty construct.** Rivalry between teams can be healthy if the core bonds of unity are not broken and if alignment throughout the organization is placed first above all else.

2. **The After-Action-Review (AAR).** The After-Action-Review is a powerful tool for teams. It allows all members of the team to communicate openly, regardless of rank or standing. The AAR results in a better understanding of what went wrong in an operation, why, and how to fix key factors so that the same mistakes are not repeated in the future. The AAR is an excellent tool for the military and business alike.

3. **Making seemingly counterintuitive decisions.** Leaders must sometimes make the tough call that may seem counterintuitive but is actually 'spot on.' Leaders must often go with their instincts honed by years of experience.

4. **It is important to remember the families of team members.** Families of team members sacrifice as much as those who serve in the military or in business.

> *The greater the loyalty of a group toward the group,*
> *the greater is the motivation among the members to*
> *achieve the goals of the group, and the greater the*
> *probability that the group will achieve its goals.*
> ~RENSIS LIKERT

CHAPTER 8

THE POWER OF PERSEVERANCE

Through perseverance, many people win success out
of what seemed destined to be certain failure.

~BENJAMIN DISRAELI

A TOUGH PERSONAL LESSON

While at Fort Hood, I learned a very powerful lesson that was best expressed by neurologist and Holocaust survivor Victor Frankel, who said, "When we are no longer able to change a situation, we are challenged to change ourselves."

It began with a short-notice trip to Kuwait.

As the 1st Cavalry Division prepared for deployment, I was asked to join in a rehearsal in Kuwait. During the long fifteen-hour plane ride from Ft. Hood to Kuwait, I worked the entire time. I didn't take a break. I didn't even get out of my seat.

After landing, I went to the gym for a workout, to run quarters on a treadmill for speed work. A few minutes into my routine, a Division Commander from another division approached me and asked if I could outrun all the Battalion Commanders in the 1st Cavalry Division. I said that I had never tried so I didn't know. The Division Commander, who was always very competitive, offered up a challenge. "Let's jump on two treadmills, you on one, and I'll be on the one next to you. Let's set the speed to a six-minute mile pace." I took up the challenge, and together we ran for a while at that speed. When we stopped, the Division Commander said, "Yes, I think you can outrun all of your Battalion Commanders because I can outrun all the Battalion Commanders in my division." It was a very good workout. Later, in Iraq, that same General and some young, fast officers, and I would run together on Sundays.

It was later that day, after the treadmill challenge, and during the rehearsal for our upcoming deployment to Iraq, that I first felt a pain in my left side. The pain persisted throughout the entire rehearsal.

The next day, when I woke up, the pain was still there. Because of the pace and intensity of my impromptu running competition the previous day, I thought that I had just strained a muscle. But the pain persisted.

On the flight home, the pain worsened. I spent the entire flight bending and flexing, to try and stretch out the left side of my upper torso hoping to get some relief.

When we landed in Chicago, I called the division surgeon and asked him to meet me in my office the next morning and to bring some Motrin because my side hurt. But I didn't make it to my office that morning. As I was driving to work, the pain was so intense, I had to pull over to the side of the road and call for an ambulance.

At the hospital, the doctors performed an exhaustive number of tests but could find nothing wrong. I was about to be discharged, when the vascular surgeon said, "General Bostick, I recommend an angiogram because it looks like your left lung is not filling to capacity." Personally, I did not think I needed an angiogram. I was athletic and in great shape. Overall, I felt good, except for the pain in my side. And given that the battery of tests had not indicated anything wrong, I believed the situation was not that serious. Still, I accepted the doctor's recommendation and was transported from Ft. Hood to a hospital in the city of Temple, which had the necessary equipment and staff to perform the procedure.

I asked the Non-Commissioned Officer who had accompanied me in the ambulance to call my wife Renee, let her know that I was in an ambulance on my way to Temple, reassure her not to worry, and to let her know she did not need to come to the hospital. But Renee was very worried and didn't listen. Before I knew it, she was right there with me at the hospital.

The doctors performed the angiogram and the findings were disturbing. They discovered that I had a pulmonary embolism. They called it deep venous thrombosis. At the time, I had never heard of a pulmonary embolism, much less deep venous thrombosis. But I soon learned that two significant factors contributed to what could have become a life-threatening situation. First, I was dehydrated. Second, the doctors informed me that because I had not moved, stood up, or walked around during the flight to Kuwait, a blood clot had formed in my leg. The clot had subsequently broken loose, traveling through my heart and into my lung. This embolism was causing the pain in my side. I was admitted to the hospital and spent several days of treatment, monitoring, and observation.

But that blood clot, although serious, was not the only thing on my mind. There was also the division's ongoing preparation for our deployment to Iraq and a funeral that required a General Officer representative for the Army.

AN EVEN TOUGHER CHALLENGE

As part of our preparation for the upcoming deployment into Iraq, we were going through a Warfighter exercise, which is a major evaluation of the division headquarters. While I was in the hospital, the Commander of the 1st Cavalry Division, who would have led this exercise, had to return to his home state because of an emergency. So, the Assistant Division Commander for Maneuver had to lead the division through the Warfighter exercise. And he did a first-rate job of leading the division.

It was during the Warfighter exercise that a funeral requirement and my own immediate situation intersected. I was faced with a challenging decision.

Our division was asked to send a General Officer to represent the Army at the funeral for a soldier from the 101st Airborne Division Command. Earlier, I had served as the General Officer representing the Army for the funeral of a fallen Lieutenant from the 4th Infantry Division who had played such an important role in the story about the unifying role of friendship.

I spoke with my doctor from my hospital bed and told him that I was checking myself out of the hospital to represent the Army at the funeral of this soldier. The doctor tried to dissuade me from leaving, but I passionately believed that others had suffered much more than I had. I thought that I could perhaps offer the family comfort and assistance with my presence. The doctor pointed out all the negatives of me leaving the hospital prematurely, but I was determined to attend that funeral.

The only thing I asked was that the doctor tell me what I had to do to travel safely to Houston.

The doctor finally relented in the face of my determination and cleared me for travel. I was ordered to wear compression socks, special long elastic stockings designed to help increase circulation and help prevent venous issues such as blood clots and potentially another pulmonary embolism. I had to ensure I was well hydrated by drinking lots of water. And I had to take breaks and walk around during the drive to Houston. I followed these directives to the letter and made it safely to Houston without any further medical complications.

The funeral I was to attend was for Specialist Ray Joseph Hutchinson who had served in the 2nd Battalion, 502nd Infantry Regiment, 101st Airborne Division (Air Assault). Specialist Hutchinson had been scheduled to return home on December 6 to be with his family for his grandmother's emergency surgery. That was the plan. And then the plan changed. Ray Joseph learned that by flying out on December 6 he would bump one of his fellow soldiers from a pre-scheduled flight home. So, Ray Joseph gave up his seat and told his squad leader that he would just wait. He would see his family a few short weeks later, at Christmas.

Tragically, the next morning, December 7, while on patrol, Ray Joseph's vehicle was struck by an Improvised Explosive Device (IED). It exploded into his vehicle. The explosion was fatal. At the age of twenty, Specialist Ray Joseph Hutchinson was killed while serving his country.

When I arrived at the church for the funeral service, I met with Ray Joseph's mother, Deborah, his father, Michael, brother Lee, and their entire extended family. I recall his parents, Deborah and Michael, both giving me tearful hugs and thanking me for attending their son's funeral. But it wasn't just me they thanked. They thanked the Army for giving Ray Joseph an opportunity to serve his country, something he had desperately wanted to do. I have remained in touch with the Hutchinson family over the years, and I have told the story of Ray Joseph Hutchinson on numerous occasions as a testament to the qualities and attributes of such a selfless servant.

After the funeral for Ray Joseph Hutchinson, I returned to Fort Hood to prepare the division for deployment. The Christmas holidays were approaching, as was our deployment date. But since I was now on the blood thinner Coumadin for six months, I knew that I would not be permitted to deploy with the 1st Cavalry Division.

Regarding the upcoming deployment, I had to make a decision that was best for the division and my family. I walked into the Division Commander's office and told him that I planned to use the upcoming holidays to transition out of my job, and it would be best for him to find a new Assistant Division Commander for Support.

Our Division Commander didn't accept the plan that I proposed. His response to my suggestion was simply, "No, you're deploying with the division." I pressed on. I pointed out, "The Surgeon General of the Army himself told me that I could not deploy." Our Division Commander knew how difficult this situation was for me, and how hard I had worked to help the division prepare to deploy. He offered a solution. "If you get a doctor's statement that asserts you can deploy, then you will deploy," he assured me.

Following my Division Commander's guidance, I prepared a letter describing my condition. I listed all the possible risks, that a cut or injury would result in excessive bleeding because of the Coumadin treatment I was still undergoing. I pointed out that there would also be the inconvenience of having my blood checked every week. But then I listed all the positives. I pointed out the precautions I would take to ensure that my condition would not limit my performance or negatively affect any part of the operation. I followed with a strong statement stressing that with the proper mitigation efforts, it would be safe for me to deploy the division into Kuwait and move the forces into Iraq. For the letter's signature block, I used the name of a doctor friend who attended the U.S. Army War College with me. This friend was an Airborne- and Ranger-qualified doctor and spent a good amount of his time in special operations. When one of the Army Corps was split between Afghanistan and Iraq, he was the officer called upon for his leadership expertise to serve as the Corps Chief of Staff in Afghanistan. He knew how to balance taking prudent risks and mission requirements.

I asked my friend if he would sign this letter. I made a full disclosure revealing that several senior leaders had already told me that I could not deploy. He signed the letter without hesitation and said, "Be safe, Tom, our doctors will take care of you in Kuwait."

With my letter in hand and still on Coumadin, I prepared to deploy the division of over 25,000 soldiers, their equipment, and myself to Kuwait.

I put the medical issue behind me, but little did I know that this medical experience still had one more important lesson that I would learn.

After returning home from training in preparation for deployment, I told my wife about a General, a colleague and friend, who always had a white Corvette parked in front of his headquarters. He was always in the field training, so that car just sat there in its parking space. It never moved. Nobody ever seemed to drive it. So, often when I saw another Corvette, I would mention this story about this General to my wife. After so many mentions of this Corvette, my wife came to believe that I really liked Corvettes.

BUT THAT CAR IS ONLY HALF THE STORY

One day, while this General with the Corvette and I were in the gym, the conversation turned to travel, especially lengthy international plane trips that lasted for twelve hours or more. And I shared with him that it was on just such a long trip to Kuwait that I developed a blood clot in my leg that had resulted in a pulmonary embolism.

You can imagine my surprise when he shared an almost identical story to mine, one that had also resulted in a pulmonary embolism. He told me that it happened on a trip from Washington, D.C. to Europe. The trip would normally have taken around eight or nine hours, but his plane sat on the tarmac for several hours before being cleared for takeoff. Those extra hours had added significantly to the total duration of the trip. His situation was further complicated by the fact that he was assigned a window seat and didn't want to disturb his fellow passengers who were sleeping in the seats next to him. He remained in his seat. He didn't stand up. He didn't walk around. He didn't move. The next day he wasn't feeling well and when he sought medical attention, the doctors informed him that he had a pulmonary embolism.

Somehow, knowing that this General, who was also very fit, had developed a pulmonary embolism, come through it with flying colors, and was now back in the fight, gave me great hope. This General demonstrated personal resilience. It gave me even more faith that I would prevail and fully recover. I told my wife about this General, and she could tell that there was clear positive change in my mindset and my emotions, and a strong belief that I would fully recover.

But that wasn't the end of the effect of this General's story, his pulmonary embolism, or his Corvette.

It was late afternoon on Christmas Eve 2004 at Ft. Hood. I was home alone. Renee and our son Joshua had been out shopping for several hours. I started thinking that it was getting late and I should call to check on them, when suddenly they appeared in the kitchen looking a bit excited. They told me that they needed my help carrying groceries and shopping bags into the house. I followed them out the door to the driveway and stopped dead in my tracks. "There's a red Corvette in the driveway," I said. "That's *your* Corvette, and it's candy apple red," Renee told me.

Then she handed a set of car keys to me. I couldn't believe it. "Why did you get me a Corvette?" I asked. Renee countered with, "Why don't you just say, 'thank you' rather than 'why did you buy me a Corvette?'" She went on to explain that I was always talking about this other General's Corvette, and how knowing that he had also suffered a pulmonary embolism and had completely recovered had given me hope. So, Renee thought a Corvette would help lift my spirits and take my mind off my own pulmonary embolism. The Corvette would be a reminder that if this other General could completely recover from a pulmonary embolism, so could I. "But Renee," I said, "I'm going to deploy for the next year." She said, "That's okay, the Corvette will be here ready and waiting when you return, just like your friend's Corvette is always there ready and waiting for him. The Corvette will just sit here for a while."

I learned a lot from my health experience. It was a challenging time for me. As a fit athlete, I wasn't used to any physical challenges, especially challenges I had brought upon myself. But my embolism gave me an understanding of the physical stresses of long-haul air travel and how to make provisions for it. I learned the importance of understanding that on a person's most challenging days, other soldiers and families have their own challenges and need support. I also learned the power of sharing something that in my mind was a negative with a trusted friend, and subsequently see what I thought was a weakness as a strength. And finally, there was that amazing candy apple red Corvette. Given my many years away from home, I finally decided to sell it. With several life lessons learned, it was time to move on to my next duty assignment.

WHAT I LEARNED

1.   **Personal focus.** Leaders must focus on their own personal resilience and pay close attention to their own mental and physical fitness by preparing their mind and their body for setbacks.
2.   **Helping others when you are down.** A leader can have a bad day, but not in front of the troops. For the troops will continue to require great leadership. Supporting others will give renewed strength to the leader. To quote Booker T. Washington, "Those who are happiest are those who do the most for others."
3.   **Selfless service.** As Ray Joseph demonstrated, many soldiers in combat are concerned most about their fellow soldiers. That is what strong teammates do. In business, leaders should consider what sacrifices teammates make and why they make them. Is the sacrifice for their fellow workers, the company, or something else?
4.   **Accept support.** When knocked down, learn to accept support from family, friends, and colleagues. In this way, leaders can find themselves winning after losing due to a health setback.

5.   **Bounce back strong.** After experiencing a setback, it is important to bounce back stronger than before with even greater mental and physical readiness as well as the skills to employ protective measures. Personal resilience is developed over time by preparing for setbacks through mental and physical training, absorbing the impact by bending but not breaking, recovering properly, and then adapting to become stronger than before.

Persistence and resilience only come from having been given the chance to work through difficult problems.

~GEVER TULLEY

CHAPTER 9

TURNING DIRT
SURPRISING STORIES OF
RECONSTRUCTION IN IRAQ

The project has not started until you're turning dirt.

~GENERAL GEORGE CASEY

THE CHALLENGE OF INTERNATIONAL TEAMWORK

Acompelling article in *The Harvard Business Review* discussing successful global teams points out that "to succeed in the global economy today, more and more companies are relying on a geographically dispersed workforce. They build teams that offer the best functional expertise from around the world, combined with deep, local knowledge of the most promising markets. They draw on the benefits of international diversity, bringing together people from many cultures with varied work experiences and different perspectives on strategic and organizational challenges. All this helps multinational companies compete in the current business environment."

While the Army isn't a global company, it is a global organization with a "geographically dispersed workforce" and it does bring together "people from many cultures and different perspectives" to work as teams.

The *Harvard Business Review* article goes on to point out some tough truths: "But managers who lead global teams are up against stiff challenges. Creating successful workgroups is hard enough when everyone is local and people share the same office space. But when team members come from different countries and functional backgrounds and are working in different locations, communication can rapidly deteriorate, misunderstanding can ensue, and cooperation can degenerate into distrust."

Three of these "tough truth" challenges surfaced during the building and managing of our teams dedicated to the mission of reconstructing Iraq.

As Casey Stengel, famed manager of both the New York Yankees and the New York Mets, once said, "Getting good players is easy. Getting 'em to play together is the hard part." Creating teams that played well together allowed them to reach their goals successfully.

#1 THE GRAVEL CHALLENGE

Who would have thought that the construction of an entire airfield would come to a total standstill for the lack of something as simple as gravel? But it did until a surprising solution came to light from a very unlikely source.

One of our more important projects was to rebuild a strategically important airfield in Iraq. This was a key project. But our construction had suddenly come to a complete standstill. Why? Because we were unable to move gravel, one of the key building materials required, from the local quarry to the construction site due to ongoing attacks against the convoys of trucks.

We had a meeting with several General Officers to try to get to the bottom of the problem. We tried to find out how to have the trucks complete the drive safely from the quarry to the airport.

There was a stretch of road between the quarry and the airport construction site that was constantly interdicted with small arms fire. The truck drivers were terrified and could not deliver the gravel along that dangerous road.

Our solution?

We made gravel transport safer in two ways.

First, we added additional security vehicles to the gravel convoys to protect them as they drove the route from the quarry to the airfield.

Second, we added police barricades at the entrance to the main road and restricted entry to only those vehicles going to the site. This was to ensure that unauthorized persons would be blocked from accessing that critical construction route and our team of drivers would be safe.

Once these protection strategies were in place, we were confident that the problem was solved. And it was. The trucks loaded with gravel started to deliver their loads to the construction site—for a while. And then they stopped again.

The next set of issues we encountered was at the quarry site itself. We discovered that our quarry teams were being attacked while they worked on excavating the gravel and loading the transport trucks.

Our solution? We added additional security for the onsite quarry workers.

Again, our efforts to protect the quarry team worked—for a while. And then they stopped working.

The problems escalated. This time our Iraqi quarry workers were not attacked on the job—it was worse than that. Our Iraqi workers were being attacked in their homes. Not only were their own lives at grave risk but by working to move gravel to the airfield construction site, they were putting their families' lives at risk. So, our Iraqi teams stopped coming to the quarry site altogether.

The gravel challenge seemed insurmountable. Until Mo.

Mo stepped up and offered a solution, that ultimately proved as effective as it was creative.

My local interpreter was named Mohamed, or Mo for short. One day, we were on our way to yet another meeting with a group of Generals hoping to find a permanent solution to the gravel problem so that our airport construction could get back on track. As we traveled, Mo started a conversation that would turn out to be the very best solution.

Mo turned to me and said, "General Bostick, I think I can help you with the gravel."

I was unconvinced and responded, "Mo, I'm meeting with all of these Generals, and we've not been able to figure out a solution, but you, my interpreter, you think you can help me?"

Mo's confidence was unshaken. It didn't matter to him that he was my interpreter or that all our Generals had been unable to find a permanent solution to the gravel challenge. "Yes, Sir, I believe I can," he responded. "Just let me go up to the quarry area and speak with the locals."

I reluctantly agreed to let him go but insisted that he be accompanied by U.S. security forces. I wasn't about to risk sending him up there on his own, completely unprotected.

Mo declined, saying, "No Sir. I need to go alone. I'll take my personal car. I have a pistol and an AK-47. I'll be okay."

Against my better judgment, I allowed Mo to go on his own.

He was gone for a couple of days. I did not sleep at all while he was away. Finally, Mo returned. He came back safe, successful, and with an innovative solution.

"Sir, I figured it out," he said. "This issue is not a problem with the enemy. It's all about the local tribal leaders. They want a piece of the action."

In just two days, Mo had tapped into one of the key components of successful team building. Our problems were all about a better understanding of the local culture. Mo understood that. He understood the local culture. He respected the traditions and authority structure of the local tribal culture. He knew how to work effectively with the local culture.

I was intrigued and asked, "So what do you suggest?"

"Sir, you know that large quarry contract that you have with the big U.S. contractor?" he asked.

"Yes, of course," I replied, wondering where this was going.

"You should break that contract into several smaller contracts and disperse them among the tribal leaders and their companies, and don't just give the entire contract to a company that isn't local," Mo announced.

I was not convinced. "Mo, we would lose the money, time, and efficiency gained by working with one large contractor."

He replied with quite a bit of perception. "Sir, how much time, money, and efficiency have you lost by not finding a solution to this problem thus far?"

He was right.

We broke up the large contract into several smaller contracts and distributed these among the tribal leaders and their organizations. The attacks stopped. The road became safe to travel. Workers were not threatened in their homes. And construction continued smoothly and without further incident. The critical airfield was completed. The project was a success. Mo was right.

Mo and I met again. This time in the United States.

Iraqi civilians who supported U.S. Forces in Iraq had the opportunity to come to the United States if supported with a letter from a U.S. General Officer. I wrote several of these letters for highly dedicated Iraqi civilians who worked for my team. Mo was one of those who asked me for such a letter and one whom I supported. Mo is now in the United States. He has been my guest and visited my home. This man accomplished so much for my team in Iraq. He knew the culture better than the other Generals and I did. He was an integral part of my international team.

#2 THE POWER GRID CHALLENGE

Restoring electricity was one of the most important missions in Iraq. One of the power plants where electricity had to be restored was a key power plant in northern Iraq. This was a very challenging project in a highly contested area. It was a dangerous place to work.

The contract for the restoration of the power plant had been awarded to a firm from the United States. The CEO and I met at my office in Baghdad to discuss the project. We were both on the same page. He spoke Portuguese as did I, so we had a common bond through another language.

The project began uneventfully. Work started and continued smoothly. There were no issues or problems. And then, a few months in, the power plant was attacked, and even worse, the project manager who led the onsite construction team was kidnapped.

The ramifications were serious. Sadly, the project manager who had been captured was never found despite an immediate and thorough search. As a result, work on the power plant came to a complete stop as the CEO pulled his team and his company off the power plant project and ultimately out of Iraq. Given the importance of this power plant, we were under intense time pressure. We had to resume construction immediately.

We had two options. Option one was to reopen the bidding process and start from scratch to find a new company to complete the work on the power plant. If we reopened the bidding process, we'd face significant delays.

Our second and better option in terms of time would be to try and reopen communications with the company CEO who had been awarded the original contract and whose construction team had already made progress in rebuilding the power plant.

Option two seemed like the better choice. We started by setting up a Video Teleconference with the CEO. Our entire team—finance, legal, and operations—came together to participate in that strategically critical call. When we connected, I said to the CEO, "We are deeply sorry for your loss, and completely understand why you pulled out. However, that doesn't change the fact that this project remains critical, and we must start work again." I then went on to say, "I am prepared to reopen the bid process on this project and find another company to award the power plant contract to, but before doing so, I wanted to ask how much it would cost for your team to return to the site." After a long pause, the CEO provided a number.

We agreed to the number on the spot. We also reconfirmed that the project team would be onsite as soon as possible and that we would have enhanced security on the site for their protection. With the decision made, the Video Teleconference ended.

That's when the finance and legal team jumped in and questioned whether it was appropriate for me to make such a unilateral decision without further analysis. I responded by asking them, "What's a project manager's life worth? And how do we put a price on completing this important project for the country of Iraq?" We proceeded with the decision.

Sometimes, more analysis and consensus in decision-making is not necessary, and leaders must go with their instincts.

The original company was back onsite with a full construction team within weeks, and the power plant was successfully restored as planned.

#3 THE "SHARE THE SUCCESS STORY" CHALLENGE

Benjamin Disraeli, a British politician who twice served as Prime Minister of the United Kingdom wrote, "Without publicity, there can be no public support, and without public support, every nation must decay." I had never imagined that I would find myself not only understanding the wisdom of Disraeli's words but developing a campaign to implement them.

What started as a simple project to share a wonderful success story that needed to be told about children and their new school, later seemed an insurmountable project as we ran into some significant obstacles.

Children are special, and schools are special. Put those two together and you have a formula for success—or so I thought.

The children in Iraq are like children everywhere in the world. They are curious, excited, trusting, easy to please, and persistent. They are filled with joy and bring joy to everyone who encounters them—soldiers from far away as well as villagers from nearby. Building schools for children in Iraq really provided our forces with important grassroots support at the local level. Children, their families, and in fact the entire village would celebrate the building of a village school. The children and the school created a bond between our soldiers and civilians living in those villages. And those bonds brought with them support.

However, that support was limited to the small circle of people who could see the completed schools. While we received great support from the people in each village with a new school, the people in the villages without a new school for their children remained skeptical that we were doing anything significant to assist the local population.

Even with media highlighting local school projects that we completed, people in other parts of the country did not trust U.S. claims of success that they could not witness for themselves firsthand. The Iraqi people needed to have confirmation that our troops had indeed built a school, and plenty of them. And they needed that confirmation to come from the local Iraqi leaders who had first-hand knowledge of the success of these projects. The fact that we were building schools for village children was simply not believed by people who were unable to see those schools and those children for themselves.

This challenge was one for which my team had a novel solution.

During a visit to a project in the Kurdistan region of northern Iraq where we had just finished a school for the local children, we knew how important it was to share this story with other regions in Iraq and show the positive effect and constructive relationships we were building through our school projects. So, I asked the mayor if he would speak about this project on a video that we were making.

His immediate response was no. He pointed out that it was dangerous to be associated with the U.S. forces in Iraq, even to talk about something as pleasant as a school for children which the U.S. had built.

Even though his fears and concerns about sharing this story of successful school construction with the rest of the country were understandable, I was disappointed.

It was important to find a way to get this story told by an Iraqi leader. Several weeks later, during the official ribbon cutting for that same school in northern Iraq, I asked the mayor if he would speak in the video if it was shown only in Kurdistan. This time his answer was, "Yes. I'll do it." That was a big win for the children, the village, and the public perception of our U.S. forces in Iraq. Again, understanding the culture made all the difference.

These three "tough truth" challenges—the gravel site, the power plant, and the school—could have resulted in three failures instead of three successes, but just as that *Harvard Review* article pointed out, it is important to be sensitive to the culture, to listen to local people who know best what's really happening on the ground, and to be patient and creative enough to develop solutions to challenges that work for everyone. All best practices for meeting challenges and getting results.

WHAT I LEARNED

1. **Cultural awareness**. It is crucial to listen to people who may know best what is happening on the ground, who are familiar with the local culture and its customs. It is then crucial to adapt your own thinking and actions to succeed in that new culture.

2. **Two kinds of team players**. In today's global economic and organizational context, teams are frequently made up of both internal organizational members as well as external players. It is important to have an inclusive team atmosphere in which external team members feel important and their expertise is valued.

3. **Go with your instincts.** Sometimes the most efficient and cost-effective solution will not succeed. Know when to change. If one effort fails, adjust, and try again. Often a leader must go with his or her instincts. Know when to pivot to an approach that may not be the most direct but may be more successful.

> *Cultural understanding is necessary both to defeat*
> *adversaries and to work successfully with allies.*
> ~WILLIAM D. WUNDERLE

CHAPTER 10

FIRST, BREAK ALL THE RULES

In recruiting, there are no good or bad
experiences—just learning experiences!

~ANON

TEAM RECRUITING

Turning things around is often a double-edged sword. On the one hand, it holds the bright promise of success; on the other hand, it holds the dark fear of failure. Turning things around all too often brings with it the very real risk that change and innovation will clash with tradition and ritual—and that tradition and ritual will win.

This unexpected situation was the challenge I was presented with in 2005 just after returning from Iraq.

And it all started with a phone call.

I had completed my assignment in Iraq and returned to Washington, D.C. to be assigned as the Deputy Chief of Engineers which was the second highest position in the U.S. Army Corps of Engineers. Over the next few months, we settled in. My wife Renee had accepted a position as principal of an excellent public elementary school, just a few miles away in Arlington, Virginia. We bought a house. Our lives fell into a comfortable, settled, and welcome routine after so much time apart while I was deployed to Iraq.

Then, one afternoon on my day off, as I was hitting golf balls at the driving range, my phone rang and everything changed.

The call was from the Chief of Engineers and Commanding General of the U.S. Army Corps of Engineers who told me to call the General Officer Management Office (GOMO) as they had been trying to reach me. That did not bode well. Something was up. And I had the feeling it would be something I didn't want to hear.

The GOMO informed me that I had been reassigned. This turn of events was difficult to believe. Having just returned from Iraq to assume the position of Deputy Chief of Engineers in Washington, D.C., how was it possible to be reassigned after just a few short months?

The GOMO was understanding and sympathetic, but the reality was the Army had a much bigger challenge and a critically urgent mission. The Army had just failed its 2005 recruiting mission. The quota of new recruits had not been met. And it didn't seem to make any difference that I had no experience in recruiting—having never been assigned as a recruiter. The Army had a problem and selected me to lead the team at U.S. Army Recruiting Command (USAREC) in solving it.

The mission was to take command of USAREC, Fort Knox, Kentucky, in two months and help "turn around" USAREC to achieve recruiting mission success.

GOMO did try to soften the blow by pointing out that the recruiting assignment was estimated to last only eighteen months and that the Army wanted me to command a maneuver division. Commanding one of the Army's ten maneuver divisions was appealing, but that did not make the move any easier.

I set to work making plans.

Since I would only be at USAREC for eighteen months, it did not make sense for my wife to quit her job, or for us to sell the house we had just purchased. We decided that we would manage, and I would be a geographical bachelor for the eighteen-month assignment at USAREC. So I packed up a few items and headed to Fort Knox, Kentucky.

But at Fort Knox there was a major surprise waiting.

WHEN COMPETITION CAN BE THE PROBLEM, NOT THE SOLUTION

The competition to fill the monthly recruitment quota was significant. USAREC had a very competitive environment with soldiers competing with other soldiers to achieve their quota of new recruits each month.

The recruiters' job was a lonely one. The traditional team culture for which the Army is known was not evident. In USAREC there are permanent recruiters who never depart from USAREC until they retire, and they are then combined with "detail" recruiters. These detail recruiters are soldiers who left combat units to serve on a "detail" or "temporary" assignment in USAREC for just a few years before returning to other units in the Army. What these detail recruiters experienced when they joined USAREC was a very different type of teamwork than what they had experienced in the rest of the Army, particularly in combat, from where many of these soldiers were just returning. Duty in USAREC was a more solitary mission rather than a team mission. Each recruiter was incentivized and rewarded for their individual achievement.

The work climate could be brutal. The workdays were long; ten-hour days were not unusual. Soldiers worked most weekends. Time off was a luxury. I recall my first visit to a Recruiting Battalion and asking the Battalion Commander to show me his time-off policy. The Battalion Commander proudly pulled the policy from his notebook and showed it to me. What I read shocked me. The policy clearly stated that "no soldier will work past 10 p.m. without his permission, and every soldier will have one weekend off per month." Reading the policy in disbelief, I looked at the Battalion Commander and said, "You're kidding me, right?" He said, "Sir, I'm the nice guy, the old policy stated that no soldier would work past midnight."

The competition was fierce. The pressure was intense. Time off from recruiting was rare. Soldiers worked endlessly to accomplish their individual recruiting mission, which was generally to sign up two new recruits per month. All soldiers tasked with signing up new recruits were totally on their own. They sank or swam pretty much based on their own natural ability and skill set. And a soldier's success or failure was tied directly to their ability to make their monthly quota.

Those were some of the key reasons Recruiting Command was not the first choice of many officers and noncommissioned officers. In fact, it was so unappealing, that many soldiers tried anything to avoid serving in Recruiting Command.

Having just returned from Iraq, I thought, "What have I gotten myself into?"

The answer: get some context on Army recruiting. In 1973, the draft ended, and the All-Volunteer Force began. However, most recruiters will say that we have an "All-Recruited Force" because very few young men and women simply "volunteer."

Clearly, the method of operation to achieve success was not working. We had a problem. Something had to change. And that something would have to fix the problem.

THE LEADER AS CHANGE AGENT

The opportunity to become that change agent came during one of my early meetings with the leadership of Recruiting

Command. During that meeting, one of the items on the agenda was a scheduled upcoming meeting at the headquarters that Saturday with the ten lowest performing Battalion Commanders. These ten Commanders were the ones whose battalions had failed to accomplish their mission of new recruits that month.

I asked, "What is the purpose of the Saturday meeting with these ten Battalion Commanders?"

A senior Non-Commissioned Officer explained, "Sir, these low performing Battalion Commanders come to the USAREC headquarters at the end of the month and explain to you, Sir, why they are doing so poorly, and how they plan on improving the situation before the next month." He added, "We've been doing this for years."

I responded, "Cancel the meeting. I will visit the Commanders at their sites, and it will not be on Saturday."

A Sergeant Major on the headquarter's staff responded, "Sir, this is not a gentleman's sport."

I replied, "Sergeant Major, I appreciate your thoughts, but this meeting is over, and you and everyone else in my office can leave now."

It was important to send a message that changes were coming to the climate and the culture of Recruiting Command. And that the first step to make those changes had already been put into action. The "we've always done it this way" mindset would have to go. It would be replaced with a fresh vision.

My next act was to move away from the solo recruiting model that had been in place for decades and wasn't working and replace it with a team model, based on the existing Army team culture.

We made the decision to set up a pilot.

We had six recruiting brigades across the country and forty-five recruiting battalions across USAREC. My next step was to call the 3rd Brigade Commander.

"We are going to start team recruiting and one of your battalions will pilot the concept."

His first question was, "What is team recruiting?"

I explained, "Currently, if there are four recruiters in a recruiting station, and each of those recruiters has a mission to bring in two new soldiers per month, then each soldier is solely responsible for figuring out a way to bring in those two recruits all on their own. We're changing that. We're changing that mission from an individual mission where each soldier was responsible for recruiting two new soldiers per month, to a team mission where a station of four recruiters will be responsible for recruiting eight soldiers per month; if there were five soldiers in the recruiting station, then the team mission would be ten. This mission will be assigned to a team rather than an individual mission."

The goal through the new team recruiting structure was designed to identify an individual recruiter's actual talent and cast him or her in a role that would leverage that talent and add it to the pool of team talent. This goal would allow the team to succeed in its mission.

Despite the stiff resistance, we moved on to implement the new recruiting model and change the negative recruitment rate to a positive one.

TESTING STARTS WITH A PILOT PROJECT

We started team recruiting in the 3rd Brigade which was located at Fort Knox. This was our pilot organization. Our incubator. A lot was riding on both the feasibility of the team model and on the 3rd Brigade—the advance guard. Their experience would create the path, if successful, for all of USAREC to follow. We took small steps. We wanted to first test our new model.

Who did we select?

The first battalion that would test the change to team recruiting was our Milwaukee Recruiting Battalion led by Lieutenant Colonel Ted Behncke. Once that battalion had been set up to work under the new team model, we would add the entire 3rd Recruiting Brigade to the new team model.

My first pep talk was designed to encourage the leaders to buy into the new model. My goal was to have them onboard enough to give this new team concept a real chance.

Later, the entire 3rd Recruiting Brigade leadership team came together in the USAREC auditorium, and we discussed how important it was to work as a team. They were reminded that the team concept was what we had all experienced throughout the Army. It was a highly successful model that had been responsible for the success of the Army in many missions. They were assured that we could replicate this same tight team model in Recruiting Command.

Many in that auditorium had their doubts, and they were beginning to make me privately doubt my own decision. I asked myself if making this massive change was the right thing to do, especially considering our failure to meet our recruitment goals in 2005? Was the timing right to completely change the way things had been done for more than thirty years? Despite my doubts, I knew deep in my heart that we had to change, and we had to change now.

At the end of the "pep talk" to the 3rd Brigade, Lieutenant Colonel Ted Behncke approached me and said, "Sir, I think I figured it out."

"Figured what out?" I asked.

"Team recruiting, Sir," he responded. "I think I know how to do it."

I remember thinking that, finally, someone other than me believes this can work.

Ted went on to lay out his thoughts.

He explained that in any given station, there are soldiers who are great communicators and some who are not. Then there are some who know exactly how to speak to young men and women, parents, friends, and school administrators. There are also the soldiers who understand that the Army has over 150 different military occupational specialties (MOS), and what that means in terms of a wide variety of opportunities that can be offered to new recruits. Then there are the soldiers who are computer savvy, who know how to close out all the paperwork necessary to bring someone into the Army. Other soldiers are great at physical fitness and understand how to prepare young men and women to enter the Army.

Knowing the importance and diversity of the necessary skill sets, Ted said he organized his recruiting stations first around the strengths of individual soldiers and then grouped them together into effective recruiting teams, where each soldier's unique skill set would enhance the performance of the entire team.

THE BEGINNING OF A TRANSFORMATION

The transformation didn't happen right away. As Ted explained, "It took some time." And then he added, "Sir, team recruiting is now working in the Milwaukee Recruiting Battalion."

My first thought when I heard this? Hallelujah, this will work!

My next thought? Call the 3rd Recruiting Brigade Commander and model team recruiting after the Milwaukee Recruiting Battalion concept.

Once we brought the entire 3rd Recruiting Brigade onboard and expanded the pilot, our next action was to bring all the leaders together for a discussion. The goal was education. As Harvard professor and acknowledged authority on leadership and change John P. Kotter points out, "One of the most common ways to overcome resistance to change is to educate people about it beforehand. Communication of ideas helps people see the need for and the logic of a change."

It was at that meeting that I gave every leader in the room the book, *First Break All the Rules, What the World's Greatest Managers Do Differently* by Marcus Buckingham.

Our "team recruiting" concept was breaking all the rules. The 3rd Brigade Command Sergeant Major described the situation best when he observed, "This is three decades of conditioning that we're trying to break."

How did we do it? Did we pass our first test?

The 3rd Brigade did not do well initially in terms of meeting their recruiting targets, but we had expected there would be challenges during the transition. Once again, we learned that change is hard. And as the 3rd Brigade Commander correctly pointed out, the biggest challenge would be in moving away from the "this is how we've always done it" mindset.

It took some time, but eventually all brigades would change to team recruiting, and for the next ten years the Army would not fail to achieve its recruiting mission.

ONE CHALLENGE OFTEN LEADS TO ANOTHER

Yet we weren't nearly done. We may have solved the recruiting problem as far as enlisting soldiers were concerned, but we also had another challenge.

The next challenge was in recruiting our medical professionals.

We didn't realize it at the time, but we also needed to change that outdated model as well. So, in addition to building powerful teams with different skills and united through a common mission, we had to push the edge of the envelope and take on the creation of a specialty-focused team—a medical recruiting team.

The first realization of the problem came when my Physician Assistant, Captain James Jones, asked for a meeting with me. James had enlisted in the Army, then entered the Officer Candidate School to become an officer. He went on to receive his Ph.D. In short, James was brilliant, energetic, and extremely innovative. One day, James came to talk to me about creating a Medical Recruiting Brigade. He explained that each of the six Recruiting Brigades had one medical recruiting battalion and five or six enlisted recruiting battalions. Understandably, their priority was to focus on the enlisted recruiting mission and not the medical recruiting mission. Enlisted soldiers are the lifeblood of the Army. James understood that the priority for the Army was the enlisted mission, but that meant that bringing new medical recruits into the Army was not a high priority, and therefore the medical recruiting battalions were allocated minimal resources. The result? The medical recruiting battalions consistently failed in achieving their mission to bring in doctors, nurses, dentists, and other medical professionals. James's solution was to create a dedicated Medical Recruiting Brigade.

Initially, I was not open to the idea.

I had grown up serving in combat divisions where we utilized the concept of a task force which included all elements of the combined arms team (infantry, armor, artillery, engineer, medical, logistics, etc.). I thought of the Enlisted Recruiting Brigade just like those maneuver brigades in our combat divisions where a task force included everyone. All elements of the maneuver brigades worked together to accomplish the mission. I didn't think that separating out the medical recruiting battalions would serve any useful purpose and might also undermine the Army team concept we were trying to emulate among all the Recruiting Brigades.

But on further examination and investigation, the concern that we were not optimizing the recruitment of medical professionals became clearer.

One of the reasons was simple. The problem was that the medical recruiting battalions simply didn't receive the attention and resources they needed.

Each quarter, we held Quarterly Training Briefings with the Brigades. However, the Brigade Commanders always had their medical recruiting battalions brief last. Unfortunately, we generally ran out of time for these medical recruiting Battalion Commanders to say very much. It was clear that the medical recruiting battalion mission did not receive much of the attention of the Enlisted Recruiting Brigade leaders.

There was also another issue.

The medical battalions received very few resources such as marketing dollars when compared to other enlisted recruiting battalions because they were considered lower priority.

The combination of poor statistics for recruiting medical professionals, little time for their Commanders to even speak in our meetings, and minimal allocation of resources made me think that James was right.

Something had to change. And that change would start with my own perception. I had to change. I had to acknowledge that there was a difference between my understanding of a task force in a maneuver unit and the "task force" in a recruiting brigade. In a maneuver task force, the entire unit was focused on one objective. However, with the Enlisted Recruiting Brigade, there were two objectives: the enlisted recruiting mission, which was the top priority for the Army, and the medical recruiting mission, which was given low priority by the enlisted recruiting brigade leaders.

We made the leap to accommodate the special circumstances of medical recruiting. The USAREC would now pull the medical recruiting battalions away from each of the six brigades and form a seventh recruiting brigade, the Medical Recruiting Brigade.

THE EVER-PRESENT RESISTANCE TO CHANGE

Again, there was strong resistance to this change.

Case in point: a superb Sergeant Major on the USAREC Staff that I had selected to be the first Command Sergeant Major of the new Medical Recruiting Brigade was not pleased with his selection. I knew that moving him to a dedicated Medical Recruiting Brigade was the right thing to do when the Sergeant Major asked me, "Sir, why are you selecting me for this job? We send the losers into medical recruiting." Those words confirmed that James was right when he first brought to my attention the fact that medical recruiting was shunned and considered a low priority.

With the decision made, what was needed now was persistence despite the unfavorable odds. We created the Medical Recruiting Brigade and, not long after, resistance faded as we demonstrated the success of the strategy by achieving the goal of the medical recruiting mission. The excellent Sergeant Major on the USAREC Staff became the Medical Recruiting Brigade Command Sergeant Major and proudly led them to success in achieving their mission.

Now, more than a decade later, the Medical Recruiting Brigade is still making a difference for USAREC and the Army.

While the new team recruitment model was beginning to show solid success, we weren't out of the woods yet. An unexpected new challenge arose in a very dramatic fashion.

And at the root of this problem was a lack of message alignment.

When I was selected as the head of U.S. Army Recruiting Command, the Army had failed to achieve its 2005 recruitment goal. For the first time in several years, the Army had not fulfilled its recruiting mission. The entire recruitment model was revamped from 2005 through 2009, and these were some of the most challenging times in recruiting. We were making system-wide changes. We were disrupting long-held traditions and methods. We were retraining over nine thousand recruiters. We were moving fast.

As if that wasn't challenging enough, we were also recruiting the soldiers to support the "surge" of troops into Iraq and Afghanistan. Young men and women who enlist in the Army today may not report to basic training for eight to ten months. During my time in recruiting, they were reporting to basic training within two weeks after signing an enlistment contract. Again, speed and sheer numbers were required.

With the speed of change, the pressure to succeed, and the vast numbers of both recruiters and recruits during this time, we had to ensure that our message to the American public was clear and accurate. However, we had communication challenges.

The severity of these communication challenges and training gaps was brought into sharp focus through a disturbing news broadcast.

One day, a featured news segment highlighted a hidden recording in one of our stations in our New York Recruiting Battalion. To say that this segment was damaging would be an understatement.

It seems that a young reporter walked into one of our recruiting stations in New York City armed with a hidden recording device. Pretending to be interested in joining the Army the reporter asked the recruiter on duty, "If I join the Army, will I have to go to Iraq?" Our recruiter replied reassuringly, "No, haven't you heard, the war's over, so you would not have to go to Iraq."

Of course, that was not a true statement. The war was not over. Iraq was a strong deployment possibility for anyone enlisting in the Army.

The fallout was significant. I was called to Congress to testify on this matter, and we were directed to immediately place recording cameras in all stations of the NYC Recruiting Battalion offices.

I felt strongly that the cameras were not the answer. Most recruiters have the highest integrity and military bearing. The few who do not would certainly not say anything inappropriate in front of cameras. We had to change the culture of how we communicated clearly and consistently across the command.

I did feel that this communication issue had highlighted not only a gap in our recruiter education program, but also opened an opportunity to strengthen the entire recruiting team and improve our recruitment outcomes even more.

It was clear that our recruiters required a consistent message, no matter whom they were talking to or where across the country those discussions were taking place. We had to make sure that everyone was on the same page.

This led to the creation of the "Pocket Talking Points" initiative. Each month USAREC published key talking points on a single sheet of paper that was distributed to every single one of our nine thousand recruiters. The single page was portable. It could be easily folded and placed into the cargo pocket of the soldier's uniform. The message across the board was the same. There were no deviations. It was understood that there would be no ad-libbing or extemporaneous messaging allowed. There was no excuse for going "off script" in terms of the high-level messages, except that recruiters could talk about opportunities for potential recruits that aligned with their passion and talent. Those pocket talking points were policy.

From then on, every time leaders visited a recruiting station, we expected soldiers to have, understand, and utilize their pocket talking points.

The solution was successful. We finally had over nine thousand recruiters communicating the same message across the command.

This solution did not necessarily mean that recruiters did not occasionally stray from the talking points, or put their own spin on the message, but we had something that we could rely on as a standard with a common message throughout the command. Speaking with one voice was essential and expected.

We had solved several challenges in USAREC. We had solved the issue of competition between individual recruiters within a station by creating a cohesive team model. We had solved the problem of the medical recruiting battalion being low priority and unsuccessful by creating a dedicated Medical Recruiting Brigade just for healthcare professionals. We had solved the problem of recruiters accidentally or intentionally misrepresenting information by creating pocket talking points that would be sent to every single recruiter every month to ensure the message was consistent across the command.

But we weren't done yet. There was one more challenge that had to be solved.

A FACT OF LEADERSHIP LIFE—THERE IS ALWAYS ONE MORE CHALLENGE TO BE SOLVED

While there were many changes that we implemented at USAREC, one that brought our communities into the recruiting fight was the concept of the Grassroots Community Advisory Board.

As part of my routine to visit recruiting units, I attended a meeting in upstate New York. A young Major from the recruiting battalion sat at the head of the table. Around the table were seated key community leaders: the Mayor, the head of the bank, the principal of the area high school, representatives from the local television and radio stations, and several others. The focus of the meeting was the upcoming 4th of July Parade and the plan to recognize veterans.

Since it was such a community-centric meeting, I was curious and asked the Mayor why the Major was running the meeting. The Mayor replied, "The Major is a great leader and the recruiting battalion is a good rallying point for the community."

That Major and that community had taken it upon themselves to create a channel for communication and support, a unique collaboration.

Civilians and different organizations across the United States often asked me how they could help the Army with its recruiting mission. This Major had given me an idea. We needed something like that 4th of July city planning committee across the board in all our major cities focused on recruiting year-round, something that would make it easier for communities to get involved and provide support.

This idea ultimately led to the creation of the Grassroots Community Advisory Board.

During my travels across the country, many community leaders would continue to ask how they could help. At the time, there was no organized method or channel that allowed these leaders to support our recruiters and the Army. But the Grassroots Community Advisory Board provided an answer that could consistently answer this question across the United States.

After visiting upstate New York and watching that Major run a community meeting, we started the Grassroots Community Advisory Board to foster interaction and support from the community in an organized fashion.

The first two pilot programs were launched in Dallas and Los Angeles.

The first Grassroots Community Advisory Board was held on June 13, 2007, in the Dallas Cowboys football stadium. Jason Witten, the Cowboys superstar tight end, was our guest speaker. Seated around the large table were high school principals, university presidents, leaders from radio and television, Civilian Aides to the Secretary of the Army (CASAs), and many others. Also represented were the Ross

Perot organization, the North Texas Chapter of the League of United Latin American Citizens, leaders from the North Texas education system, and the Mayor of Fort Worth.

In Los Angeles, Tommy Lasorda from the Dodgers hosted our meeting with similar community leaders. My deputy at the time, Brigadier General Joe Anderson (and later, Lieutenant General Joe Anderson) was the speaker at these initial events. And two of my long-time personal friends, Eric Nishizawa and David Iwata, were involved with the Los Angeles Grassroots Community Advisory Board from the beginning.

It wasn't long before we had Grassroots Community Advisory Boards for every battalion.

And more than a decade later, the model for community support that we created is still "in the fight" helping with the local recruiting mission as they continue to support U.S. Army Recruiting today.

Another challenge resolved. Another positive solution. And another success for the U. S. Army Recruiting Command.

Closely related to the Grassroots Community Advisory Boards was the move to include spouses of troops in suggesting new strategies and solutions.

We are very aware that our mission is to not only recruit a soldier but also to retain a family.

Having learned that troops closest to the action often have the best view of the changes that might make a difference, we held the Commanding General's All-Star Advisory Council in November 2006. Of the nearly nine thousand recruiters, we brought together the 125 most successful recruiters along with their spouses. Spouses also serve their country by supporting their soldier. During the advisory council meeting, recruiters and their spouses provided more than two dozen recommendations for us to consider. Some of these recommendations would help lead USAREC to success in the years ahead.

But all good things must come to an end.

My "eighteen-month assignment to turn things around" turned into nearly four years as the head of recruiting. But during those four years, we made positive changes that are still making a difference in USAREC today.

It is important to also acknowledge that not every change works as envisioned. And that was also true for me in the recruiting command.

Since I remained at Recruiting Command for four years, this extended assignment also provided me with the opportunity to realize that some of our changes did not work. I had a front-row seat not only for the ideas that succeeded but also for those that failed.

I had asked my Deputy, Brigadier General Joe Anderson to take a hard look at some of the changes we made during the last four years. I wanted him to assess those changes that were working and identify those changes that were not working. And his analysis shed a light on some of the problems that our changes created rather than solved.

For example, to maximize the efficiency of our resources, we centralized all the human resources (HR) at the headquarters in Fort Knox, Kentucky, and reduced the overall number of HR personnel. This change of centralizing HR personnel was particularly challenging for the brigades. We subsequently realized that HR is a resource that was more effective if it remained at the local level where local Commanders and personnel specialists had a better understanding of the people. So, we admitted that this change of centralized HR was not working, and we moved HR assets back to the local brigades.

We had tremendous support from the Secretary of the Army, the Army Chief of Staff, and the Army Vice Chief of Staff. They provided the additional resources to help identify sticky problems and provide long term solutions. We had the ability to use a talented firm like McKinsey to standardize and document the Grassroots Advisory Boards. McKinsey also helped to change from a philosophy of bringing the lowest performing Commanders to the headquarter on Saturday to a coaching and mentoring performance review, which was important and rooted the change permanently in the culture.

We benefited greatly from the support from the Office of Economic and Manpower Analysis (OEMA) at West Point. OEMA is part of the Social Sciences Department at West Point, and they provided a tremendous amount of support in terms of the efforts that were working and those that were not working in order to recruit the right men and women into the Army. The team at OEMA included some of the Army's brightest and most creative minds, and they truly made a difference for the Army's manpower efforts.

Finally, we knew it was time to move USAREC into the twenty-first century in terms of information technology. When I first walked into a recruiting station the recruiting map was an actual paper map, with clear plastic overlays and push pins to identify key locations. We converted these recruiting stations into twenty-first century war room operations centers with big screen televisions, digital maps, and data-driven details that would adapt based on the changing demographic environment where they met. In some cases, we were ahead of our time. For example, we requested permission to create a Facebook account for each recruiting battalion. That request was denied, but today all units in the Army have their own Facebook account.

During my tenure, we solved many problems. And in the process, we learned a wealth of valuable lessons.

WHAT I LEARNED

1. **The best time to change is often just after being defeated.** The key to winning after losing is recognizing that this is often the right time to move in a new direction regardless of how challenging it may be. It is important not to stay with a bad decision but to put it behind you and create a new path.

2. **Be creative in the composition of your team.** Think outside the organization when you are building your teams.

The Grassroots Community Advisory Board reinforced for me that the Army team is broader, including not only uniformed soldiers but also patriotic civilians.

3. **Speak with one voice.** It is important when leading a team or an entire organization to ensure that your message is consistent and internalized. Effective leaders and everyone on their team must speak with one voice.

4. **Listen to your people in the trenches.** Those junior people in the trenches, who must execute policies, could very well know best how to succeed in their unique environments and assignments. Keep an open mind.

5. **Be intentional.** Go see for yourself. Ask questions. Listen. Make mental notes. Take written notes. Celebrate success.

6. **Set aside time to reflect and assess changes that have been made.** Recognize when the change envisioned, planned, and executed is not working as originally desired and don't be afraid to admit defeat and change again. Change is constant.

Life is short. Break the rules.
~MARK TWAIN

CHAPTER 11

OUR SOLDIERS

A hero is an ordinary individual who finds the strength to persevere and endure in spite of overwhelming obstacles.

~CHRISTOPHER REEVE

THREE STORIES

Often it is not the people that make the headlines who are the heroes. More often than not heroes are those everyday people who inspire, persevere, and endure. They all have dreams. And it is how they overcome these challenges to realize their dreams that is cause for celebration.

Each of these stories showcase three individuals who, against all odds, reached for their own personal dream.

While serving as the Commanding General of U.S. Army Recruiting Command, I had the opportunity to travel all over the country as well as internationally to visit with our great soldiers, civilians, and families involved in the all-important mission of recruiting the men and women to serve in our Army. During these travels, many of these great Americans who decided to serve their country shared their stories. Also apparent during these trips was the outpouring of support for our soldiers as well as for the civilians and families who were directly involved in protecting our country.

During those times when recruiting is challenging, there is always the concern that the Army will lower its standards and accept soldiers of lesser "quality." The stories of these soldiers, however, demonstrate that the quality by which we may judge an individual should go well beyond the common metrics of education, test scores, current fitness, age, or even some errors in judgment they may have made in the past.

Statistics are important, but they are not always the best indicators of potential. Many statistics predict that some individuals who have faced challenges in life may not be successful in the military. However, it is essential to remember that statistics don't always tell the whole story.

What matters and what our recruiters attempt to determine is not only the past history of an individual, but his or her future potential. It's the potential of an individual, and their ability to meet the demands, the values, and the life of the American soldier which count.

There are success stories that should give us all a reason to question our assumptions. These stories offer us powerful proof that we should all encourage our young men and women to explore the opportunity of wearing the uniform of the United States military.

Quietly, and without fanfare, many of these soldiers contribute greatly to our nation's security. With less than one percent of Americans serving in uniform, these soldiers are generally unknown to most Americans. And yet, their stories are a testament to how they faced significant challenges and adversity, how each persevered, and ultimately achieved success.

The story of three such soldiers follow. These soldiers understood the concept of winning after losing. Their stories reflect selfless service, enormous sacrifice, and pride in serving the Nation.

STORY #1 RODERICK EVANS

The greater the obstacle, the more glory in overcoming it.
~MOLIÈRE

One of the major health concerns in the United States is obesity, especially among our youth. The military represents a cross-section of America, so the young people that we attempt to recruit today have many of the same challenges as many other Americans, including weight issues.

One of these recruits was Roderick Evans. He was passionate about serving his country, but his first Army recruiter told him that he was too heavy. A self-proclaimed "Snickerholic," Roderick decided to lose the weight necessary to join the Army. How much did he weigh? Roderick weighed 418 pounds. Roderick's goal for many would seem an insurmountable obstacle. After all, millions of Americans struggle with weight, try to lose it, and fail. But Roderick had a driving goal. He wanted to be a soldier. That was his dream. And that dream was powerful enough that Roderick was inspired to lose 230 pounds—more than half his body weight—to enlist in the Army.

Obesity is a national security risk because it limits the available pool of fully qualified young men and women, a pool that continues to decrease because Americans are getting heavier. Statistics show that obesity has more than doubled since the 1980s. As the Army works to recruit young men and women, it is faced with real challenges to recruit from a population where one in four children are obese.

When I served in Recruiting Command, over 33 percent of Army recruits came from the southeastern United States. While obesity has increased across the country, obesity in the southeastern part of the United States is among the highest in the Nation. Obesity has nearly tripled in young people under the age of eighteen in just twenty-eight years. Between 16 and 33 percent of adolescents are now obese, such that in certain parts of the United States an alarming one in three children is obese.

Obese children are at risk for numerous health challenges, diseases, and medical issues, including type-II diabetes, cardiovascular disease, and hypertension. Diabetes is the seventh-leading cause of death in the United States. With between 100,000 and 400,000 obesity-related deaths per year and an estimated healthcare expenditure of $117 billion, obesity has surpassed health-care costs related to smoking and drinking. High cholesterol, high blood pressure, joint problems, sleep apnea, and psychological or social challenges are just the beginning of what an obese child could potentially expect in their lifetime.

Thus, the story of Roderick Evans is a bright spot in this pervasive issue of obesity. During my visits around the country, I often spoke about Roderick Evans and the amazing commitment he made to lose almost half his body weight so he could join the Army.

Roderick's story was so compelling that it had to be shared with all the recruiting leaders at our annual training conference. However, rather than me telling his story, I decided it would be best if Roderick told his story himself. We located him in San Antonio where he was serving as an Army medic. Before long he had joined our recruiting leaders at our annual training conference and impressed us with his incredible story of how he overcame one of the hardest personal barriers of all—obesity—to achieve his desire to serve the Army and the Nation.

Roderick still carries a picture of himself at 5 feet 7 inches, 418 pounds as a reminder of his past and the "never quit" attitude he had to adopt to succeed.

But weight wasn't Roderick's only challenge.

At the time he first applied he was not only over the weight limit for admittance to the Army, but he was also over the age limit, which was set at thirty-five years of age, by law.

Even though he had this other seemingly insurmountable challenge of age restriction hanging over him, Roderick was neither dissuaded from his course of action or discouraged during his three-year crusade of eating healthier smaller portions, first walking and then running to lose weight. He knew that even though he might lose the weight and qualify for enlistment into the Army, he would never be able to push back the clock and lose years.

None of that mattered to him. Roderick fought a tough personal battle to secure his place in the Army. He pursued his dream, one day at a time.

And he won. He won both his weight battle and his age battle.

In 2006, Congress approved a new Army policy which allowed men and women to enlist up to the age of forty-two. This new enlistment age limit policy removed the last barrier between Roderick Evans, now thirty-nine years old and a lean 165 pounds, from his dream.

What was he like in person? During our first meeting, he seemed very quiet. But when he took the stage in front of all my recruiting leaders, he was electric—a gifted, inspirational, and motivational speaker.

Was there anything in his childhood that would have given any indication of his extraordinary determination?

Not really. He grew up in Detroit with a single mother who worked two jobs in order to care for her four children. He started college in Louisiana at Grambling University and went on to work as a trombone instructor and vocal teacher for the next fourteen years. When Roderick's brother became sick, he returned to Michigan with his wife to help his brother. Roderick was always a very caring person, and this attitude served him well in the Army starting at basic training where he was a natural leader. The younger soldiers affectionally and with great respect called him "pops."

Roderick served as a 91W, a medic. He was a role model for soldiers who looked to him for inspiration. But he hasn't forgotten his roots. He often returns to his hometown of Detroit to speak and to inspire youngsters to never give up on their dreams, to persevere, to endure.

STORY #2 JACKIE PURRINGTON

When you get into a tight place, and everything goes against you till it seems as if you couldn't hold on a minute longer, never give up then, for that's just the place and time that the tide'll turn.

~HARRIET BEECHER STOWE

Jackie Purrington and I first met in Iraq in the summer of 2004. She was a project manager engineer and was deployed as a part of the many Army Corps of Engineer civilians serving in Iraq and Afghanistan.

That first meeting was memorable. Jackie was a new civilian member of our team, and as a civilian, she also wore a military uniform without any rank or branch insignia. However, she was a bit distraught because a Command Sergeant Major, the Non-Commissioned Officer whose role it was to maintain the standard of wearing the uniform properly, had stopped her as she walked to my headquarters. He had given her a difficult time because of the way she was wearing her uniform. Her pockets were open. Her bootlaces were not tucked into her boots. But her worst uniform transgression was that she had decided to remove her desert camouflage uniform (DCU) shirt because the temperature was well over a hundred degrees. Without wearing her camouflage uniform shirt, the Command Sergeant Major let her know that she was completely out of uniform with her T-shirt and Army trousers. This initial experience with the Command Sergeant Major was a difficult start for Jackie.

But despite her rocky beginning, Jackie turned out to be an excellent program manager.

She got her chance to shine when she was asked to replace a senior officer with about eighteen years of Army experience and who had been serving as assistant to the Lieutenant General responsible for developing, organizing, training, equipping, and sustaining the Iraqi Security Ministries and the Iraqi Security Forces. She accepted the assignment.

I saw her new boss, the Lieutenant General, in the gym the next morning. When he approached me, my first thought was that he might not accept Jackie because she was not a soldier. As it turned out, it was the opposite. He told me that Jackie was doing a fantastic job. In fact, she did such a fantastic job that her stay in Iraq was extended twice.

Ironically, the Command Sergeant Major who had first called Jackie out for improperly wearing the uniform was the senior Non-Commissioned Officer for this organization that trained Iraqi soldiers. Jackie would be serving every day with him, but now she completely understood how to wear the uniform.

In the end, it was that very uniform that changed Jackie's life and demonstrated her remarkable degree of determination. It happened this way.

When it was time for Jackie to return home, I asked Jackie if there was anything that I could do for her. She quickly said yes. She wanted to join the Army and asked for my help.

But there was a problem. Jackie was over the age limit for joining the Army, which at that time was thirty-five. I pointed out that she was already making an extraordinary contribution to the defense of the Nation by serving in Iraq as a civilian—not once but twice.

But for Jackie that wasn't enough. She wanted to wear the uniform as a soldier, not as a civilian. She had trained hard to be accepted. She exercised. She was very fit. She served side-by-side with soldiers every day. But most of all she loved the camaraderie of the military. She wanted to be part of it. She wanted to be a soldier.

I told her that I would follow-up on her request with U.S. Army Recruiting Command (USAREC). I called the Commanding General of USAREC on Jackie's behalf. I wasn't surprised by his response. He told me that there was no possible way that Jackie could join the Army. Her age was her biggest hurdle and one which only an act of Congress itself could remove.

And that's exactly what happened.

Not long after I became the Commanding General of USAREC, Congress approved a higher age limit for new soldiers. The new age limit was now forty-two.

Jackie was ecstatic. On August 18, 2007, at forty-two years of age, she was commissioned 2nd Lieutenant Jackie Purrington, an Engineer Officer.

That wasn't Jackie's only first. She also became the first officer to receive a direct commission as an Engineer Officer through U.S. Army Recruiting Command.

Jackie would later redeploy to Iraq, proudly wearing the uniform of an Engineer Officer. She made it. She was a soldier!

STORY #3 ANGELO VACCARO

Perseverance is a great element of success. If you only knock long enough and loud enough at the gate, you are sure to wake up somebody.
~HENRY WADSWORTH LONGFELLOW

Weight and age were not the only challenges that some individuals had to overcome. Some potential recruits, like

Angelo Vaccaro, had behavior issues. As a teenager, he ran with a rough crowd in the Bronx but never got into serious trouble himself. However, with the way his life was going it was only a matter of time.

Then fate stepped in. The Vaccaro family moved to Deltona, Florida, and Angelo had an opportunity to reinvent himself— to change his life. And he grabbed it.

Angelo wanted a better life, so he changed his behavior. He enlisted in the Army. He became a health care specialist assigned to Alpha Company, 1st Battalion, 32nd Infantry Regiment, 3rd Brigade Combat Team, 10th Mountain Division.

From that moment on, he dedicated his life and his service to saving lives.

Corporal Vaccaro had developed a history of assisting and rescuing soldiers under heavy fire. On July 5, 2006, he rescued three wounded soldiers, carrying them one-by-one down a 7,500-foot ridgeline while under enemy fire. He shielded those soldiers with his own body even though he himself was suffering from shrapnel wounds. He had come a long way from his days as a teenager in the Bronx.

Then in 2006, on October 3, the twenty-three-year-old Angelo made the ultimate sacrifice.

He was attempting to recover two-platoon mates wounded in a firefight in the Korengal Valley of Afghanistan when a rocket-propelled grenade hit and killed him. Angelo didn't have to go on that mission. He was still recovering from the injuries he had sustained just three months earlier.

His brother, Ray, credits the military with Angelo's transformation, saying how he was changed from a lost youth to a passionate combat medic proud of helping people heal. Angelo might have started as a young man destined for a life filled with challenge, but he ended life as a hero.

In 2007, a building that housed a Warrior Transition Brigade at Walter Reed Army Medical Center was named in honor of Corporal Vaccaro. During the dedication ceremony, the Commander of the Walter Reed Army Medical Center told the assembled audience how appropriate it was to name the building after Angelo, who overcame delinquency to become a decorated combat medic and whose actions embodied the Army values.

The plaque on that building is inscribed: "October 3rd, 2006, Corporal Vaccaro learned that his platoon was involved in direct fire engagement with enemy forces and he volunteered to conduct an extremely dangerous ground extraction."

At that time, Corporal Vaccaro was the only double Silver Star recipient in the current conflicts. The Silver Star is the third-highest personal decoration for valor in combat just behind the Medal of Honor and the Distinguished Service Cross. Corporal Vaccaro received his first Silver Star for action in July of 2006. He received his second Silver Star posthumously for his fearless efforts in the Korengal Valley of Afghanistan. Corporal Vaccaro demonstrated the heroic behavior seen only in the most courageous of people. He did indeed represent the best of our soldiers.

WHAT I LEARNED

1. **Never give up.** The stories of courage and tenacity in this chapter show how important it is for leaders to instill into their teams the idea of never giving up on your dreams. A great deal can be achieved by those who work in an environment of support and encouragement.

2. **Go as far as you must go to make positive change happen.** Sometimes you must surmount many obstacles but always keep moving forward. Go as far as you need to go to find your dream, to make a difference, and to effect change whether in the military, in business, or government.

3. **Heroes are often ordinary people doing extraordinary things.** In each of these stories, it was the personal drive and belief in themselves that helped each of these heroes achieve his or her dreams. Will power, perseverance, and support from others made all the difference. These soldiers demonstrated significant personal resilience.

People do not decide to become extraordinary.
They decide to accomplish extraordinary things.
~EDMUND HILLARY

CHAPTER 12

PANCAKES-FOR-DINNER STRATEGY

When you join the military, you join
the largest family in the country.

~ANON

A FAMILY ALSO SERVES

My father was a soldier. My mother was a soldier's wife. My family was a soldier's family. It was during my years growing up that I came to understand the importance of family in military life. I learned that when a soldier enlists to serve his or her country, the whole family enlists and serves right alongside, making a significant difference in a soldier's career.

FROM ICY UNIFORMS TO SPIT-SHINED SHOES

I first became aware of the bond between a soldier and family when I noticed the extraordinary care my mother took with my father's uniforms. As soon as they were out of the wash, my mother always put them, soggy and wet, right into the refrigerator. This extraordinary measure helped her iron these starched uniforms so that my dad looked sharp for work every day.

But it wasn't only my mother who helped.

Each one of my siblings—Michael, Kathy, Anthony, and Peter—made their own unique contribution to my father's professional, confident, military look. My job was to shine my father's boots each night. I became particularly good at spit shining boots. First, a layer of polish was applied, and the boots were set aside to dry. Then they were buffed to a mirror finish with a soft polishing cloth and a little bit of spit. It wasn't easy and it took time, but spit shining meant my father could see his face reflected in the shine of his boots.

That skill served me well years later when I arrived at West Point. One of the key challenges that plebes (freshmen) had was spit shining their shoes. Given my experience with my father's boots, I was able to help several of my classmates look sharp.

SUPPORT FROM THE EXTREMES

Spouses and children are not paid by the Army for their work, but their efforts clearly make a difference to the success of their soldiers.

I have always learned lessons by understanding what is happening at the extremes. Support from spouses, family members, and friends can make a significant difference, not only for serving soldiers, but also for the staff in every organization from the leaders to the managers, to the hundreds of thousands of workers in the ranks. I personally witnessed the support from spouses, family members, and friends countless times during my career. Three such stories of support that really stand out for me follow. These stories may be at the extremes of providing support, but they each demonstrate the sacrifices made and the support provided by families.

STORY #1: 172 BIRTHDAY CAKES

Our family did not have much from a financial perspective, but we did have a lot of love growing up. My father was an enlisted soldier and my mother stayed home to take care of all of us. I was already at West Point and on my way. But my parents still had four other children they had to help through school. They made sure we had what we needed, but there was little or no money for extras such as travel. So, during my four years as a cadet at West Point, I never had a visit from anyone in my family. Not until it really counted—my graduation—and then my mother traveled across the country from California to West Point, New York to see me graduate. My mother did not like to fly, but she was determined to see me graduate from West Point, and she did. My father had to continue to work and support the children and to help pay for my mother's trip to West Point, so he could not visit. However, both my father and mother were able to take the short drive from Monterey to Palo Alto to see me graduate from Stanford.

My parents ensured that all five of their children had the opportunity to attend college, and all five us became college graduates. My older brother, Mike, attended UC Santa Barbara to study art and is a gifted artist. My sister, Kathy, attended UC Santa Clara and then UC Berkeley Boalt Law School. She worked for then U.S. Attorney Rudy Giuliani in Manhattan and later for U.S. Attorney Robert Muller in San Francisco. My younger brother, Anthony, would become an All-American wrestler at UC Davis. He would later earn his veterinary medicine degree from Tuskegee University and serve for nearly thirty years as an Army Veterinarian. My youngest brother, Peter, after his studies at UCLA, became one of the leading surgical oncologists in the country.

And no matter how hard they worked or how many responsibilities they had, my mother and father always remembered special times. Every single birthday, they sent a birthday cake for me to enjoy and share with my classmates at West Point. It was then that I realized how important a birthday cake could be, especially when you are far away from home.

AND THAT'S HOW THE BIRTHDAY CAKES TRADITION WAS BORN

My first assignment in the Army was at Wildflecken, Germany. Wildflecken was a few kilometers from the former East

German border and thousands of miles away from home for my soldiers.

I recalled how I felt every year at West Point when my birthday cake arrived—feeling that I was remembered on my special day, that there were people who cared about me, and that I was not forgotten.

I decided that I would add birthday celebrations to the traditions for my platoon. We would celebrate birthdays—all the birthdays in my platoon.

Each month, I would purchase a cake and add the names of every soldier who had a birthday that month. These birthday cake celebrations became important almost from the very first cake. It was not only an acknowledgment of each soldier, but when everyone shared and celebrated together, it strengthened our platoon's team spirit.

At first, this tradition was not too challenging since I only had thirty soldiers in my platoon. Later when I was a Company Commander, it became quite a challenge to provide a cake and recognition for every one of the nearly two hundred soldiers under my command.

That's when the power of family made itself felt in a very big way.

I had recently married my wife, Renee, and shared with her the birthday cake tradition started with my platoon. I explained that the sheer number and significant additional cost could make it very challenging if not impossible to continue this tradition.

Not for Renee.

She immediately embraced the tradition and announced that she would make a double layer chocolate cake for each soldier's birthday.

I agreed, with one stipulation: that she could make these cakes for the soldiers only if they were not in any kind of trouble in the company.

Renee carried on with the cake tradition while also teaching second grade. My job was to ensure that my calendar accurately listed all upcoming birthdays, and to pass that information on to Renee.

The cake baking got off to a great start. There was even a day when there were three birthdays, which meant three birthday cakes for Renee to bake. No problem. Renee baked and decorated the cakes. My driver, who also played on our softball team, assisted me in carrying the cakes into the building. The soldiers were delighted to receive their cakes. The tradition continued.

It seemed that nearly every week there was at least one cake or more to carry into the office, but we never missed a birthday cake for a deserving soldier—and then one day we did.

I was reviewing my calendar and realized that I had forgotten to remind Renee about a birthday the previous week. Of all of our soldiers, I had forgotten about my driver's birthday. I couldn't believe it. He was such an amazing soldier—especially helping to deliver all those birthday cakes to all those soldiers.

Again, family proved indispensable.

When I told Renee about my oversight, she immediately went to work and baked not one, but two cakes for me to give to my driver. He was so happy to receive his cakes. But I was curious. I asked why he didn't say something to me when I missed his birthday cake. My driver said he thought that perhaps he had done something wrong and was therefore undeserving of a cake. His answer reminded me of the importance of being mindful of troops who, like my driver, are modest, unassuming, and uncomplaining. These are the kind of troops that need someone looking out for them.

Moving from cakes to flowers.

Renee didn't stop her contribution with birthday cakes. She thought the troops would like flower boxes for their barracks' windows. Flower boxes were a European custom, and we were living in Germany. When Renee shared her flower box idea, I wasn't on board at first. I pointed out that the soldiers were very tough, and I wasn't convinced that they would pay for flower boxes for their windows.

I was wrong.

Renee and the other spouses went to work and made sure the price was as low as possible and before long, there were flower boxes in nearly every barracks window throughout the company.

FAMILY MORALE BUILDERS MAKE CHAMPIONS

From softball to trips throughout Europe to homemade chocolate birthday cakes to flower boxes in the windows, all these gestures of caring leadership went a long way in building morale and fostering excellence. Supporting that leadership were the spouses and other family members who made these extra touches happen.

As a result, not only did this company win the U.S. Army Europe Softball Championship, but they were number one in the Army in Maintenance and the runners up in Europe for the Itschner Award given to the Best Engineer Company in the Army as discussed earlier in this book. Our 2nd Platoon received the V Corps Distinguished Small Unit Award, and Bravo Company slaughtered our opponents during the 54th Engineer Battalion Organizational Day where family members also competed.

Once again, family and friends can have a significant impact on the success and morale of any organization, whether in the military, government, or the private sector.

STORY #2: PANCAKES FOR DINNER

Recruiting is a challenging mission in the military, and I made it a point to meet with our most outstanding recruiters to learn what made them so successful.

During one of my visits to the Recruiting School at Ft. Jackson, South Carolina, I met with a former recruiter who had earned the title of Recruiter of the Year for the U.S. Army when he was a soldier. Now he was one of our top civilian instructors, and I wanted to learn his method for such extraordinary success in recruiting.

Our meeting was lengthy, and it was getting late in the day. He had to go home even though we hadn't finished our discussion.

As he was leaving, he turned to me and asked, "Sir, would you like to come to my house for dinner?"

I replied, "Sure, and then we can continue the conversation," thinking that would be a good opportunity to finish up our shortened meeting.

He asked, "Sir, do you like pancakes?"

I said, "Sure, I can eat anything, but am I joining you for dinner or breakfast?"

He replied, "Sir, it's dinner, as I mentioned earlier."

I asked curiously, "Pancakes for dinner?"

"Yes, it's pancakes for dinner," he affirmed.

I had never had pancakes for dinner, but I looked forward to joining this former U.S. Army Recruiter of the Year and his family for pancakes.

I arrived for dinner, and he and his wife welcomed me warmly to their home. Pancakes were cooking on the stove and they smelled great. I met their son who had just come in from cutting the grass. The atmosphere of their home was warm and welcoming.

WHAT'S THE SECRET?

At dinner, between forkfuls of the delicious pancakes, I shared the challenging time we were having with recruiting.

"What's your secret?" I asked. "How did you become such a successful recruiter during your time in the Army?"

Before he could answer my question, his wife responded. "Recruiting in our home is always a family affair," she said.

"How so?" I asked.

She was happy to share their formula for success.

"Do you know how many people my husband would have to speak to in order to bring one person into the Army?"

I replied, "Yes, it's about 150 separate engagements to bring one person into the Army."

"Here's our secret," she explained. "I call all the women. Women like talking to women. I call all the women and chat with them to determine who is interested and who isn't. I connect those women who expressed interest with my husband. He then does the recruiting."

She went on to say, "I cook, run our home, and I call the women to prescreen their interest. My son cuts the grass and does the dishes and other chores to keep our little household running smoothly. My husband focuses only on recruiting. He has no distractions. He knows that his family is behind him one hundred percent. His recruiting is a family job. That's how he became the U.S. Army Recruiter of the Year."

THE OFTEN UNSUNG AND UNACKNOWLEDGED SUPPORT OF FAMILIES

Not every organization's leader may have a "Pancakes for Dinner" opportunity to learn what spouses and families do. But there is no doubt that people who are successful at their jobs and excel above all expectations receive support from their family and friends who are not on the payroll. And organizations, if they tap into this additional resource, will benefit from families and friends in the accomplishment of their organizational goals. That's why it is important to understand, value, and recognize those who make a difference, whether directly or indirectly.

STORY #3: ARMY STRONG FAMILIES

There is great power in a motto. The few words that make up the slogan can inspire the culture, atmosphere, and goals of an organization. And the Army is an organization that offers an excellent example of the power of a motto. As retired four-star General Norman Schwarzkopf once observed, "As young West Point cadets, our motto was 'duty, honor, country.' But it was in the field, from the rice paddies of Southeast Asia to the sands of the Middle East, that I learned that motto's fullest meaning. There I saw gallant young Americans of every race, creed, and background fight, and sometimes die, for 'duty, honor, and their country.'"

It was during my time at U.S. Army Recruiting Command that we made a change to the Army slogan.

Some leaders had expressed the view that the existing slogan "Army of One" was contrary to the concept of teamwork.

We worked with U.S. Army Accessions Command (our higher headquarters) and McCann Worldgroup in New York City, in addition to several other stakeholder groups, before the Army settled on "Army Strong."

Army Accessions Command and McCann Worldgroup would roll out "Army Strong" at the U.S. Army Recruiting Annual Leadership Conference. During the launch, the representative from McCann told the leaders who were present that to be "Army Strong" you had to be *in* the Army. The thinking was that while parents made their children strong, it was the Army that would make them "Army Strong."

This seemed to make sense; however, my wife Renee leaned over to me and asked, "Are families and spouses not Army Strong?"

Having served nearly three decades at that time as an Army spouse and moving our son in and out of a host of schools, Renee was determined to expand the motto to include spouses and their children—the entire family of every soldier should be "Army Strong."

During the break, Renee approached the McCann representative and explained her logic for spouses and families being Army Strong. I also mentioned to the McCann representative that we encourage all of the Army recruiters to understand and value that "we recruit a soldier, but we retain families," so it does sound like spouses and families are also Army Strong even though they do not wear the uniform.

Ultimately, McCann discussed the concept of Army Strong spouses and families with the Army leadership which decided, yes, spouses and families are indeed Army Strong.

Renee's input was invaluable and assisted the Army in making Army Strong an even broader concept to include those who help their soldiers succeed. Families are indeed Army Strong.

WHAT I LEARNED

1. **Recognize the wider circle of contributors.** Family and friends play a significant role in the successful retention of the members of an organization. Support from spouses, family members, and friends can make a difference for most organizations. Family and friends are often part of what makes an organization strong and can influence whether a member remains with or leaves the organization. Consider how your organization recognizes these important members of the team who are not on the payroll.

2. **Pay attention to the quiet team members.** Leaders need to pay attention to the humble and quiet members of their teams. As they say, "The squeaky wheel gets the grease." Leaders must ensure they are mindful of those in the organization who are modest, unassuming, and uncomplaining. These people may not advocate for themselves but expect that their leaders will.

> *The strength of a family, like the strength of an army, is in its loyalty to each other.*
> ~MARIO PUZO

CHAPTER 13

THE POWER OF TEAM DIVERSITY

Strength lies in differences not in similarities.

~STEPHEN COVEY

E very Memorial Day provides an opportunity for
Americans to recognize and express profound gratitude
for the bravery and courage of those members of the Armed
Forces who paid the ultimate sacrifice in defense and support
of our country.

These servicemembers who came from many diverse
cultures had one thing in common—they fought and died
for their country.

Today the U.S. Army is one extraordinary team. Today's
Army represents a cross-section of our country, a kaleidoscope

of cultures, languages, and religions. Its strength lies in the bonds its members forged as a result of teamwork, duty, and mission. But that wasn't always the case.

Many diverse cultures contributed to creating the powerhouse we know today as the U.S. Army. We see their contributions at every turn.

Native Americans brought their warrior spirit, culture, symbols, and names like the Tomahawk, Black Hawk, and the Thunderbird, which became identified with missiles, helicopters, and jets in the military.

Japanese-American soldiers fought with a team called the 442nd "Go for Broke" Regiment that became one of the most decorated units in American history. They fought as part of a greater team despite their families being confined at internment camps during the Second World War.

African-American soldiers fought for America in the Revolutionary War, the War of 1812, the Mexican-American War, the Civil War, the Spanish-American War, the two World Wars—always in a segregated Army. In 1948, President Truman signed into law the Executive Order that ended segregation and created one team with diverse members.

Women contributed valuable service as nurses and spies in the Civil War. During World War II, women enlisted in the Women's Army Corps often becoming prisoners of war, as well as receiving medals and citations for their valor. In 1948, President Truman signed the Women's Armed Services Integration Act, and women officially became part of the U.S.

military. Then, in 2013 women could serve in combat units and now were integrated into the entire U.S. military team.

I had the opportunity to celebrate with two of those teams and their stories follow.

OUR JAPANESE-AMERICAN SOLDIERS

On Memorial Day 2009, I had the honor of paying a special tribute to Japanese-American Military members who fought honorably for our nation's freedom in World War II—while their own freedom and the freedom of their families were denied.

In our Army, we talk about the Warrior Ethos. It is an ethos that states, "I will always place the mission first, I will never quit, I will never accept defeat, and I will never leave a fallen comrade." Although we use the words of the Warrior Ethos more often today, the concept of never leaving a fallen comrade behind is not new.

This Warrior Ethos is powerfully illustrated in a story of two soldiers and the legendary "Lost Battalion" of the Second World War.

One of the most ferocious battles of World War II was fought in late October 1944 by the Japanese-American 442nd Regimental Combat Team in the Vosges Mountains of eastern France.

It was a rescue mission.

Two hundred and seventy-eight men of the famed 1st Battalion, 141st Infantry Regiment, the "Lost Battalion" as it later became known, were trapped behind enemy lines. When Hitler was informed, he ordered that the entire unit be annihilated. His message was that these soldiers would not be permitted to fight on what was then occupied German soil. The German forces were relentless. They attacked the stranded soldiers again and again. And with each attack, the 141st Infantry Regiment lost more and more members of its team.

There had been several attempts at a rescue by other units, but each rescue mission had failed. And then the 442nd was ordered to launch one more rescue attempt.

It was now late October. The weather was cold and rainy. Conditions were miserable. But the 442nd made up of Japanese-American soldiers was undeterred. For five days they fought day and night. And then, on the fifth day they succeeded, reached the stranded men, and saved all two hundred and eleven of the men who had survived the carnage. The Japanese-American soldiers of the 442nd did not leave a fallen comrade behind. Their team exemplified the true meaning of the Warrior Ethos.

With this story as background, I was honored when my friend Terry Shima, a member of the 442nd Regimental Combat Team, asked me to speak on Memorial Day 2009 at Arlington Cemetery.

I was doubly honored when we were able to bring together two of the veterans who had been in France, under fire on that deadly October in 1944—Astro Tortolano of the stranded 1st Battalion, 141st Regiment, and Minoru Nagaoka of the 442nd

Regimental Combat Team that undertook the successful rescue mission.

But this one act of bravery was not the only one. Japanese-American soldiers, initially part of the 100th Infantry Battalion, were absorbed into the 442nd Regiment Combat Team, the "Go for Broke" team that became one of the most decorated units in U.S. military history. The soldiers of the 442nd earned more than 18,000 decorations, including more than 4,000 purple hearts for the 4,349 wounded and killed in action, 4,000 bronze stars, 271 silver stars, 29 Distinguished Service Crosses, 21 Medals of Honor, and in less than a month of fighting, they also earned five Presidential Unit Citations. Soldiers who served in the 442nd continue to earn medals and honors to this day for their past heroism. President Harry Truman reviewed the 442nd Regiment Combat Team when it returned from Italy on July 15, 1946 at the Ellipse. This was the first time a U.S. President reviewed an Army contingent of the size of a Regiment Combat Team.

In a ceremony honoring over 33,000 Japanese-American soldiers, President Clinton said, "As sons set off to war, so many mothers and fathers told them . . . live if you can, die if you must, but fight always with honor, and never bring shame on your family or your country," adding that "rarely has a nation been so well served by a people it so ill-treated."

These heroes' stories evoke inspiring patriotism, sacrifice, and courage. Their legacy continues to demonstrate to this day the great American ideals of liberty and equality for all.

Terry and I would work together again on several important projects in the years ahead. And one such project would have profound importance and a very special place in Army history.

At the time, I was Director of Personnel for the Army. My duties included organizing the Boards to review combat medals, including the Medal of Honor as well as ensuring recognition of those groups of soldiers who may not have been properly honored for their achievements in the past.

It was during this assignment as the Director of Personnel for the Army that Terry contacted me. He wanted to secure a Congressional Gold Medal for the Japanese-American Nisei. Japanese-American Nisei are second-generation Americans or Canadians who were born in the United States or Canada but whose parents had emigrated from Japan.

The Congressional Gold Medal is the most prestigious award given to people from all walks of life. It is bestowed by the United States Congress for significant achievements and contributions to the Nation. On this occasion, the U.S. Army conducted a review that resulted in forty 442nd soldiers who did not receive the Bronze Star medal during the war. General Odierno, the Chief of Staff of the Army, and I were honored to make the presentation to twenty-two 442nd veterans who attended the ceremonies in Washington, D.C.

In 2010—after many months of tireless work by Terry, the Japanese-American Veterans, and the U.S. Army—Congress approved the Congressional Gold Medal to honor Japanese-Americans who served in combat. The Japanese-American veterans who were so recognized included soldiers from the 100th Infantry Battalion, the 442nd Regimental Combat Team, and the Military Intelligence Service.

Given my Japanese heritage because of my mother, it was such an honor to engage with the wonderful members of the 442nd Regimental Combat Team families and friends. African-American heritage from my father has also been a source of strength for me and leads into this next story.

TUSKEGEE AIRMEN

Over the years, I have had the opportunity to support and engage with the Tuskegee Airmen.

One such occasion arose on October 21, 2017, when I had the honor of supporting the Tri-State Chapter Tuskegee Airmen Scholarship Foundation Dinner at the New York University Law School.

The Tuskegee Airmen Scholarship Foundation honors the legacy of the Tuskegee Airmen by providing motivation and support to deserving students pursuing careers in aviation, aerospace, STEM (Science, Technology, Engineering, and Mathematics), and other areas.

Two Tuskegee Airmen who spoke that day shared how much they had wanted to serve their country and expressed their pride in serving with the Tuskegee Airmen. They were Wilfred DeFour, age ninety-eight, and William Johnson, age ninety-one, the baby of the group of speakers. Even though these Tuskegee Airmen and others, like my father, had served in a segregated Army, they were still proud to serve. Both veterans were pleased to be recognized and appreciated for their service. Not only had they served and fought for our country during World War II, but also in so doing, they had also fought against racism and bigotry.

While the history of the Tuskegee Airmen is short—they served from 1941 to 1946—their record is impressive. In those five years, as a team, they earned three Distinguished Unit Citations, eight Purple Hearts, fourteen Bronze Stars, and ninety-six Distinguished Flying Crosses. They clearly added great value to the military, which was reflected in their exemplary performance.

On July 26, 1948, President Harry S. Truman signed Executive Order 9981. This Executive Order would finally integrate our military. Its intent: to ensure that everyone would be treated fairly based on the merits of their performance, rather than the color of their skin. The Tuskegee Airmen and the 442nd were both reviewed by President Truman at the White House on separate occasions. To each of these legendary units, President Truman mentioned that they fought the enemy abroad but faced racism at home. This view of how the Tuskegee Airmen and the 442nd were treated must have factored into the issuance of the Executive Order that would integrate the military.

The Tuskegee Airmen Scholarship Foundation celebration wasn't the only occasion that I had to engage with the Tuskegee Airmen. There would be several other opportunities to visit with these heroes from the Greatest Generation. Additionally, in 2014, I was honored by the invitation to deliver the Tuskegee University commencement address—a great honor and one which was given to Michelle Obama the following year.

One of my key messages to the graduating students was the same message that Malcolm X had shared years before—that education was their passport to the future. It was their education that would provide opportunities and access to a world of promise and possibility.

WOMEN AT WEST POINT AND IN THE ARMY

During my first two years at West Point, there were no female cadets. The Superintendent of the Academy at the time made it very clear that women should not attend West Point. He believed that women were not strong enough and could not serve in combat units.

But his views were not shared by the military leaders, and the following year in 1976, the first class of 119 women joined the Long Gray Line—the Corps of Cadets at West Point.

Addressing the graduating class of 2016, Vice President Joe Biden said, "Having men and women together on the battlefield is an incredible asset, particularly when they're asked to lead teams in parts of the world with fundamentally different expectations and norms."

While I was a cadet, it was very clear that the first couple of classes of women had a very difficult time, but the majority would persevere and graduate. Women would continue to demonstrate that they not only belonged at West Point but that they would be among the very best of cadets. And they went on to accomplish great things in the Army. Some graduated at the top of their class, some would become the Brigade First Captain (the top cadet representing the Corps of Cadets), some would be selected as Rhodes Scholars, All-American athletes, and later serve as General Officers.

One "first" that gives me great personal pride is the first-ever women's powerlifting team. While in the Army, I was selected as a member of the 1983 All-Army Power Lifting Team and competed at the 165-pound weight class. Several years later, when I returned to West Point to teach mechanical engineering, I also decided to coach the powerlifting team. Our team was all male. Once I realized that we were excluding a key group of team members, I reflected on the words of our former Superintendent who had insisted, when I was a cadet, that women were not strong enough and could not serve in combat units.

My assistant coach (Paul Christopher) and I started recruiting female athletes from other sports and managed to put together the first women's powerlifting team. Initially, we did not compete very well, in fact, we lost our matches regularly. But in time the team improved. The members got stronger. When I departed West Point for my next assignment, Paul Christopher became the coach of the power lifting teams. We were right to launch the women's team. Just four short years after we first formed the team, the West Point Women's Power Lifting Team won the National Championship.

The history of diversity in the U.S. Army has evolved over time, and diversity is truly the strength of the Army and our Nation. That no matter your race, gender, culture, language, sexual orientation, or religious belief, you can serve our Nation. And this evolution serves to remind us that the freedoms we all enjoy should never be taken for granted. As Elmer Davis eloquently stated, "The nation will remain the land of the free so long as it is the home of the brave."

WHAT I LEARNED

1. **Recognizing our heroes.** It is never too late to recognize heroes. It is never too late to address the challenges of the past.
2. **Diversity.** There is great strength in diversity of gender, ethnicity, religion, and ideas. Diversity makes the military and civilian organizations stronger.
3. **What counts in assessing talent.** Performance is the key metric.

> *When we listen and celebrate what is both common and*
> *different, we become a wiser, more inclusive,*
> *and better organization.*
> ~PAT WADORS, HEAD OF HR AT LINKEDIN

CHAPTER 14

AT THE INTERSECTION OF LEADERSHIP AND PERFORMANCE

The pessimist complains about the wind. The optimist expects it to change. The leader adjusts the sails.

~JOHN MAXWELL

There are many components to creating a successful team, but perhaps one of the most difficult is maintaining that delicate balance between team leadership and team performance.

The success of a team is measured by its achievement of a stated goal, a goal that is achieved through the performance of each team member.

And yet, one of the most critical factors of team performance is the leadership of that team.

It is the team leader who sets performance standards and goals. And in order to lead a team to performance success, a leader is the one who makes the tough team decisions, charts the course, bears the brunt of criticism, and adjusts the sails to steer into the winds of change without capsizing.

Several times during my career, I found myself steering my teams through winds of change. And it was at these intersections of leadership and performance that I faced some of the most challenging periods of my career and learned significant team building and leadership lessons.

THE "DON'T ASK, DON'T TELL" CHALLENGE

One of the most challenging leadership assignments of my military career occurred during the time I served as the U.S. Army Deputy Chief of Staff for Personnel. In this role, I was responsible for the Human Resources policies to guide the largest and most important teams ever created, over one million soldiers and 330,000 U.S. Army civilian personnel. This was a team whose performance was unparalleled.

The human resources challenges were significant. We were faced with the challenge of transitioning soldiers returning from combat into new civilian careers; with health challenges resulting from increasing rates of suicide, alcohol and substance abuse, and cases of sexual assault and harassment; and with the sensitive challenge of religious accommodations. But one of the most challenging issues we addressed was the repeal of the law known as "Don't Ask, Don't Tell."

"Don't Ask, Don't Tell" refers to the official U.S. policy as it relates to the service of gay, lesbian, and bisexual people in the U.S. military. Instituted in 1993, the policy prohibited openly homosexual or bisexual persons from enlisting in the U.S. military, stating that they "would create an unacceptable risk to the high standards of morale, good order and discipline, and unit cohesion that are the essence of military capability." The policy further prohibited members already serving in the U.S. military who were homosexual, bisexual, and lesbian from disclosing their sexual orientation.

The policy was the subject of much controversy. And it was during this sensitive period, when the policy was under review, that I was responsible for human resources policy for the U.S. Army. This position would put my leadership squarely at the center of the shifting cultural and social climate.

My involvement began the day I received a call from the Vice Chief of Staff of the Army, or the "Vice," who asked me to represent the Army on the Comprehensive Review Group. The Comprehensive Review Group was tasked with reviewing the possible impact of the repeal of the "Don't Ask, Don't Tell" law in existence at the time. I pointed out that my civilian deputy was already serving on the Comprehensive Review Group. But the Vice was adamant. He said, "We need an Army General on the Comprehensive Review Group." I offered to place a two-star General who worked for me on the Comprehensive Review Group. The Vice said that the Army General on the Comprehensive

Review Group had to be a three-star. So, I joined the Comprehensive Review Group.

On March 2, 2010, the Secretary of Defense Robert Gates appointed the Honorable Jeh Johnson and General Carter Ham as the Co-Chairs of the Comprehensive Review Group.

The Comprehensive Review Group had a deadline of December 1, 2010, to issue their report. We worked tirelessly over the next ten-month period, always highly aware that performance was not only based on the members of a team, but on the leadership that built, managed, and inspired those teams, especially during times of change—not just change in the circumstances on the battlefield, but change to the social fabric in which the whole team operated.

Our research was global as well as national. We examined the macro and the micro components of the question we had been tasked to address. The Comprehensive Review Group studied the experience of other countries. Our research engaged every part of our military. We visited the military academies, every one of the military service components, troops both overseas and within the United States, combat, and non-combat units.

Of the many visits I made during this time, I remember two most clearly. The first, a visit to a Navy SEAL team (Sea, Air, and Land) brought home to me the reality of the intersection between leadership and performance. The second, a visit to a group based in Stuttgart, Germany, almost derailed my military career.

THE SEALS AT NORFOLK, VIRGINIA

One of my many visits on behalf of the Comprehensive Review Group was to a group of Navy SEALs stationed at Norfolk, Virginia. We met with a small group of fewer than a dozen SEALs consisting of a mix of enlisted and officers. I explained our mandate: to review the possible impact of the repeal of the law commonly referred to as "Don't Ask, Don't Tell." I also explained that our review would include input from all branches of the military, including the SEALs. One particularly big and muscular SEAL sat directly across from me and didn't seem very interested in the discussion.

Then he offered his opinion on the repeal of the "Don't Ask, Don't Tell" policy, suggesting, "If they repeal the law, and the gays and lesbians can serve openly, put them in the administration or logistics jobs. Don't put them with the killers."

Then an officer at the other end of the table said, "What about . . ." and mentioned a name.

I asked, "So you have a gay SEAL?"

The group nodded their heads, "Yes."

I looked at the SEAL in front of me and asked, "So, how's that working?"

His response was surprising.

"It's working just fine," he admitted. "He has shot a lot of bad guys right here—pointing to his forehead. He does not bother us. And we don't bother him. He just kills the bad guys."

I dug deeper. "So, in this case, you're telling me that it's all about performance and not sexual orientation, correct?"

The SEAL answered, "Yes, it's all about performance."

Then I challenged him. "So, they don't all need to serve in the administration and logistics jobs as you suggested earlier?"

He thought it over and admitted, "Yes, I guess so."

I have always felt that we learn a lot about any situation by studying what happens at the extremes. The Comprehensive Review Group had examined the potential repeal of the "Don't Ask, Don't Tell" policy with the elite SEALs who certainly operated at the extremes in terms of combat missions. What they demonstrated was that ultimately the ability of men and women to serve successfully has nothing to do with sexual orientation. The point that it's all about performance was a powerful leadership lesson.

THE STUTTGART GERMANY GROUP

During his time as the Co-Chair of the Comprehensive Review Group, General Carter Ham was also the Commanding General of Africa Command. His headquarters was in Stuttgart, Germany. General Ham asked me to represent him when the Comprehensive Review Group visited Stuttgart. At that meeting, I was joined by the Assistant Secretary of the Air Force for Manpower and Reserve Affairs and others. We spoke to a group of about five hundred joint service members.

At one point during the meeting, a man seated at the back of the auditorium raised his hand and asked several questions, one of which would trigger a completely unforeseen risk to my career.

The question he asked: "We know there is no room in the military for racists. There is no room in the military for sexists. If the law 'Don't Ask, Don't Tell' is repealed, will there be room in the military for people like me who are religiously opposed to the gay lifestyle?"

I responded with what I believed to be a balanced and accurate response that addressed each of his points, "We know that there are some racists in the military; some of them have acted on their racists views and have received appropriate punishment for these actions. We also know that there are some sexists in the military who may not want female leaders overseeing them, and some have acted out on these beliefs. They too have been punished accordingly. If the law is repealed and you maintain good order and discipline regardless of your views on the gay lifestyle, there will be room for you in our military."

I had no idea at the time what would happen to my balanced and accurate response.

Shortly after the Stuttgart meeting, the remarks I made were published out of context along with several false claims, in an editorial titled, "The New Gay Army."

The potential appeal of "Don't Ask, Don't Tell" was a divisive issue, and the *Washington Times* editorial falsely quoted me claiming that I had said, "Unfortunately, we have a minority of service members who are still racists and bigoted, and you will never be able to get rid of all of them."

The editorial went on to charge, "It's unseemly for a senior officer to equate those who hold traditional values with racists and bigots. Lt. Gen. Bostick's careless words demonstrate his unsuitability to the task, and, for that reason, he should withdraw from further involvement in The Pentagon panel set to issue a report on the new policy by Dec. 1." The point of my response was to say that if the law were repealed, and the person did not act on his religious opposition to gays in the military, of course there would be room for him. My point on racists and bigots was to agree with the person who asked the question that there is no room for people like this in our military. Yet, these racists and bigots do exist, and those that act out on their views are punished accordingly.

What was not reported was that all of us on the Comprehensive Review Group remained impartial in our views of whether the law should be repealed. None of us ever took a position until the report was near completion. At that time Jeh Johnson and General Ham would ask for our personal view on the repeal of the "Don't Ask, Don't Tell" policy. Therefore, I had not expressed a personal position on the subject in Stuttgart or anywhere else that we visited. I suspect the editorial was another effort to discredit the work of the Comprehensive Review Group by making false claims against one of its members—me.

That editorial would come up again.

I recall watching the confirmation of the Commandant of the Marine Corps. If confirmed, he would be the top Marine. The Marine General and I were from the same part of California, so I was interested in his confirmation hearing.

During the hearing, Senator Sessions, who would later become President Trump's first Attorney General, held up the editorial and asked, "Now, General Bostick says that he did not say what is written in this newspaper, but if he did, should he be a leader in the Army?" The Marine General was not aware of the article and therefore could not address the question.

The effects of the media piece were not fading. If anything, they were growing.

Even though I had released a statement denying the allegations in the editorial, it wasn't enough. The piece sparked an investigation. The Secretary of the Army, John McHugh, called me to his office and asked if I said what was written in the editorial. "Of course not," I replied, at which point I was told that there would have to be an investigation and that the Secretary of Defense, Robert Gates, would direct his Inspector General to conduct the investigation.

As soon as I learned about the pending investigation regarding my remarks at Stuttgart, I engaged Jeh Johnson and General Ham. I told them that they should just release me from the Comprehensive Review Group and not let this issue potentially undermine the great work that the team had been doing. Both Jeh Johnson and General Ham said that I would remain on the team. They would wait and see just how the investigation unfolded. I valued their vote of confidence because even leaders need support during difficult times.

On the day the investigation began, we had guests for dinner at our home. I tried to stay focused on entertaining our guests, but it was challenging. Then, in the middle of our dinner party, the phone rang. It was the Under Secretary of the Army, commonly referred to as the Under. He called to give me a pep talk, to tell me that he was confident that this matter would be resolved, and I would be exonerated. The Under had known me for a long time, ever since I was a Colonel and working as the Executive Officer for the Chief of Staff of the Army. This second vote of confidence was even more appreciated during this personally difficult time.

The investigation was thorough. And the saving grace turned out to be the procedures that had been put in place by the leadership of the Comprehensive Review Group. The way our engagements with troops had been conducted was a significant factor in the outcome of the investigation. We had not allowed tape recorders during our ten months of engagements in order to allow our troops to speak freely during the discussions about the possible repeal of "Don't Ask, Don't Tell."

While we did not tape our engagement sessions, we did take notes. Our meticulous notes, like those of a stenographer, showed nothing even remotely like the quotes that had run in the newspaper. The Inspector General cleared me of any wrongdoing.

Shortly after, the Comprehensive Review Group completed its report and made its final recommendations to the Secretary of Defense that the "Don't Ask, Don't Tell" law be repealed. The Comprehensive Review Group believed that with the proper leadership and training the repeal of the "Don't Ask,

Don't Tell" law would not have a negative impact on the good order and discipline of our military.

On November 30, 2010, the Comprehensive Review Working Group released the "Don't Ask, Don't Tell" report. And a month later, on December 22, 2010, President Obama signed the repeal into law.

To this day, there has been no subsequent negative impact on the military forces related to the repeal of the "Don't Ask, Don't Tell" law.

The successful repeal of the law without any impact to the military provided a compelling example of the success that a team can achieve when team leadership meets team performance.

THE GENDER CHALLENGE

While the Comprehensive Review Group was an extraordinary opportunity and experience, it was an additional duty for me. I had many pressing matters related to my responsibilities with the human resources priorities for the Army.

One of these was the review of the role of women in the Army.

In 2010, the Army completed a three-year cyclic review of women in the Army. Three years earlier, the 2007 review indicated a need to open greater opportunities for women, but we had not made significant gains during the three years following the 2007 review. The 2010 review reiterated the same need for greater opportunities for women.

I engaged some of my senior leader colleagues and discussed the need to review opening additional opportunities for women. Some of my colleagues believed that the Army had enough on its plate with our need to review "Don't ask, Don't Tell," that we did not need to also take on the challenge of opening any significant additional opportunities for women. I disagreed. Contrary to my opinion, we did not pursue opening significant additional opportunities for women that year. It wouldn't be until January 24, 2013, that Secretary of Defense Leon Panetta would remove the military's ban on female service members serving in combat roles, and that under the leadership of the Army Chief of Staff, the Army and other services would open all combat arms to women. That was a significant and long-awaited change that provided significant opportunities for women in the Army.

Since that time, women have proven to be highly valuable in serving in these expanded roles and provide another example of the power of the crossroads where team leadership meets team performance.

THE SIKH CHALLENGE

One of my meetings at The Pentagon was with the last Sikh serving in the Army at that time. He was with his attorney, and they wanted to discuss the possibility of additional Sikhs serving in the Army after his retirement. They also pointed out that they planned to sue the Army and me personally if we did not allow Sikhs to serve in the U.S. Army.

The Sikh religion was founded in the fifteenth century, and it is distinct from Islam and Hinduism. Sikhs have unshorn hair, beards, and mustaches, and they wear turbans to cover their long hair. Sikhs had been able to serve in the U.S. Army since World War I while adhering to their religious beliefs. However, the discontinuation of exemptions in uniform standards made it difficult for Sikhs to serve. Thus, the Sikh in my office that day and his lawyer wanted to let me know that they would take the U.S. Army to court so that Sikhs could serve in the U.S. Army while being faithful to their religious beliefs.

The Army Chief of Staff directed that we ask Training and Doctrine Command (TRADOC) for their views on the ability of the Army to train Sikh soldiers who received a religious accommodation and at the same time maintain good order and discipline in the force.

I asked the Commander responsible for basic training within TRADOC for his thoughts. After reviewing this matter, TRADOC sent me a letter stating that the Army could successfully train Sikh soldiers and maintain good order and discipline.

But this decision was not a recommendation; it was simply a response to my question about the ability of the Army to train Sikhs at basic training without disrupting good order and discipline. There were other factors to consider.

Several leaders in the military were against the idea. There was a strong belief by some leaders that Sikhs must shave their beards and not be allowed to wear a turban. Those against allowing Sikhs to enter the Army with these differences in uniform standards believed this change would result in a negative impact on good order and discipline.

After much debate across the services, the first Sikh in nearly three decades (wearing his beard and turban) became an enlisted soldier in the U.S. Army and entered basic training at Fort Jackson, South Carolina. This Sikh, who had a master's degree in industrial engineering from New York University, met all the enlistment requirements. Ultimately, he was so successful that he was given the honor of carrying the unit guidon in front of the formation during graduation. The guidon is a flag that represents the unit and its commanding officer. Carrying the guidon for a unit is a high honor.

After this Sikh entered the U.S. Army, several other Sikhs followed. However, the U.S. Army began to harden its position against admitting Sikhs. There was still a firm belief that to maintain good order and discipline a Sikh should look like a soldier—clean-shaven and no turban.

By this time, I had been appointed by President Obama and confirmed by the Senate to become the Chief of Engineers and the Commanding General of the U.S. Army Corps of Engineers. However, the Sikh challenge followed me to my new position.

It started with a call from the Prime Power Battalion Commander who told me that a captain and West Point graduate had arrived at the battalion to sign in; and he was in a full beard and wearing a turban. What was unusual about this newly arrived Sikh captain was the fact that when he was at West Point, even as a Sikh, he shaved and did not wear a turban. That was the policy at the time for him to remain at West Point, and he had complied. Now, several years later, he wanted to go back to adhering to his religious customs. He chose to start in the one tactical unit in the U.S. Army Corps of Engineers, the Prime Power Battalion. His decision drew me into the debate again. I received calls from Army leaders who were concerned about the possible approval of a religious accommodation that I might grant for this officer. I challenged these requests by asking, "On what basis should I disapprove?" We had no data showing that Sikhs wearing turbans and beards would impact good order and discipline. I called a good friend, a retired Major General and lawyer in the Army Judge Advocate General Corps. I recall this friend asking me if the good order and discipline of the Army could be negatively impacted by a beard and turban. "The Army is bigger than a soldier wearing a turban and a beard," he said. I agreed, and approved the religious accommodation for this officer. He proudly and quite effectively served in our Prime Power Battalion. Later, West Point would allow Sikh cadets to wear turbans and beards and ultimately graduate and proudly serve in the U.S. Army.

THOUGHTS AND OBSERVATIONS

These three challenges, the repeal of the "Don't Ask, Don't Tell" law, the acceptance of females in combat roles, and allowing Sikhs to serve with religious accommodations for their beard and turban highlighted the tug-of-war that all too often accompanies change. There is a fear that change, especially significant social and cultural change, will have a negative impact on the successful performance of high performing, close-knit teams. There is also the uncertainty that leadership may be unable to balance change with performance.

In these situations, the concerns proved unjustified. In fact, leadership met the challenges of change and the team still grew in strength, thrived, and excelled in their performance. The leadership proved that change and risk are not negative forces. John Kenneth Galbraith expressed it best: "All of the great leaders have had one characteristic in common: it was the willingness to confront unequivocally the major anxiety of their people in their time. This and not much else is the essence of leadership."

WHAT I LEARNED

1. **Supportive leadership.** When you know a leader has your back, you are stronger, more confident, and ultimately more successful.

2. **The leader also needs support.** It can be very lonely at the top of any organization, especially during challenging times, even a leader needs support. Reaching out to a leader by way of a simple phone call will go a long way.

3. **The importance of listening and gathering data.** Any significant change for an organization will have people on both sides of an issue. Listening, gathering information, and reviewing data and evidence are invaluable in the decision-making process.

4. **Diversity is strength.** The strength of the military or any organization is in its diversity.

5. **It is all about performance.** Rather than focusing on what distinguishes different groups, focusing on performance will be a unifying factor in building great teams.

6. **Take on challenges when you feel least prepared.** Sometimes taking on challenging matters should occur when the demands on an organization are the greatest.

Effective leadership is not about making speeches or being liked; leadership is defined by results, not attributes.

~PETER DRUCKER

CHAPTER 15

THE ART AND SCIENCE OF IDENTIFYING FUTURE LEADERS

The true meaning of life is to plant trees under
whose shade you do not expect to sit.
~NELSON HENDERSON

WHAT MAKES A FUTURE LEADER?

Part of my life's work has been to reach out and support future leaders. But in writing this book, I thought it fitting to begin with my own experience and how I benefited from mentorship. I have been fortunate in my career to serve with leaders who recognized leadership qualities in me and who taught me how to recognize leadership qualities in others. This outreach and mentorship of future leaders are essential. They provide the necessary bricks and mortar required to build a thriving organization, especially during disruptive times.

In looking back over my own career, I clearly see the faces of those who not only pointed the way for me but also shone a light on each step of my journey in the Army.

And I ask myself, what did my mentor see in me, that teen-age high school student? When I reflect on my career, I reflect on the characteristics that my own champions and mentors saw in me. I found I also sought these same characteristics in the soldiers and civilians who I identified as leaders, then selected and mentored.

Based on my experience, what follows are seven attributes of future leadership excellence. These attributes are not always easy to find, to identify, or to cultivate.

Identifying these attributes and developing future leaders became a lifelong mission.

#1 START WITH YOURSELF

Before you can identify leadership qualities in others, look inward and find them in yourself.

In my case, I was possibly the least likely candidate for any kind of future leadership role. I was one of five children. My father was an enlisted soldier and my mother was a stay-at-home mom. Money was tight. But that wasn't all. My family moved all over the United States, Germany, and subsequently from Japan to California during my junior year of high school when I was suddenly dropped into an entirely new and different environment.

A big family, little money, and a whole new world to navigate and fit into. These were significant obstacles for a teenage boy. I immersed myself in academics and athletics which were always my strengths. I loved math, science, and studying Spanish. I was on the football, wrestling, and baseball teams. My competitive spirit helped me to overcome obstacles.

#2 FIND A ROLE MODEL

My earliest role models were my father and mother. They were both very humble, caring, and hardworking people who valued education. They would sacrifice anything to support their children. Both of my parents graduated from high school, but neither had the opportunity to go to college; they were determined that all five children would have that opportunity. My older brother, Mike, was the first one in our family to go to college. We were all so proud of Mike, of his persistence, and determination. We all wanted Mike to succeed. I worked three jobs in high school and turned over most of my earnings to my parents so Mike could remain at the University of California at Santa Barbara and graduate. I also knew that I wanted to go to college. I was next in line.

#3 FIND YOUR PASSION

Finding your passion is easier said than done. My father served as an inspiration because he was a military man. My father spoke little about the Army or his experiences. We moved so many times that serving as a soldier was not something that I initially considered. I did not want to put the burden of paying for college on the shoulders of my parents and other siblings. So I thought about applying to the United States Military Academy at West Point or the Air Force Academy—both options would provide full tuition along with room and board. While my passion was not initially to serve in the military, I later found that I loved serving in the Army. My passion for the military grew over time, though it was not something that was clear to me early in my career.

#4 BE PERSISTENT

I discovered that an important requirement of gaining acceptance to either West Point or the Air Force Academy was to obtain a Congressional nomination.

I was too young to understand what Aristotle meant when he wrote, "We are what we repeatedly do. Excellence, then, is not an act, but a habit." And I was certainly too young to understand that I was about to do exactly what Aristotle taught.

I now knew that I needed that Congressional nomination. My first step was to reach out to my local Congressman. Unfortunately, my application for a Congressional nomination was denied. Next, I reached out to my state's senior U.S. senator in California asking for a Congressional nomination. I was denied. I then reached out to my state's junior U.S. senator with the same request. Denied again. It was time to think outside the box. My dad was a soldier and he had joined the Army while growing up in Brooklyn, New York. I decided to write to Congresswoman Shirley Chisolm, whose district included Brooklyn. She wrote back telling me that I had to be a resident of Brooklyn to receive a Congressional nomination from her. This was my fourth rejection.

So, having exhausted all my options, I finally gave up and decided that I would attend Monterey Peninsula Community College and become a carpenter. I always enjoyed woodworking and building.

#5 BE READY FOR AN UNEXPECTED MENTOR

I was about to learn one of the most important leadership lessons of my career. Little did I know that failures are often the seeds of future success.

An unexpected visit from Retired Brigadier General George D. Wahl, who was born in 1895 and graduated with the West Point class of 1917. I had never met him before. I had no idea who he was. And I had no idea how he had even heard about me and my many attempts at trying to get a Congressional nomination.

But somehow, he knew my story.

When we met, General Wahl said, "I heard that you want to attend West Point, is that right?"

I said, "Yes, Sir, but without a Congressional nomination, I am going another route. I'm going to the Monterey Peninsula Junior College to learn how to become a carpenter."

That's when General Wahl offered me another path—a path to a nomination. He told me that I should apply for a Presidential Nomination.

I found that suggestion completely illogical. I challenged him, asking, "General, why would the President consider me for a nomination if I could not obtain a Congressional nomination?"

He told me that the President had one hundred nominations specifically earmarked for the children of military servicemembers.

Though I wasn't optimistic, I applied.

To my great surprise, I did receive a Presidential Nomination and was immediately accepted into both West Point and the Air Force Academy. I had made it—thanks to the guidance and support of an unexpected mentor.

General Wahl, a man whom I did not know and who did not know me, took the time to find me, encourage me, and mentor me with advice that changed my life forever. I graduated from West Point in the Class of 1978. General Wahl died less than three years later in 1981.

General Wahl planted a "tree" which was me, a young seventeen-year-old whom he helped to gain admission to West Point. General Wahl never knew that the "tree" he planted would grow to become a three-star General and the Commanding General of the U.S. Army Corps of Engineers. General Wahl has been an enduring influence in my life, and because of him, I have made it my personal mission to assist young men and women who may need my help.

Always be ready for that unexpected mentor.

#6 BECOME A ROLE MODEL

This is the story of a boy, his father, a cadet . . . and a three-star General that began with a chance encounter. I was attending the ceremony for the Black Engineer of the Year in Washington, D.C., when an elementary school student and his father approached me and asked if they could take a picture with me.

After the photo, the father looked at his son and said, "Tell the General what you want to be when you grow up."

The boy looked up at me and replied, "I want to be a soldier."

"You do not need to say that because I'm here in uniform," I said.

The young boy replied, "I didn't say that because of you."

"Well, then why do you want to be a soldier?" I asked, intrigued.

The boy pointed to a young West Point cadet quite a distance away and said firmly, "I want to be like him."

I asked the young boy and his father to wait for a minute. I walked over to the cadet. "Do you know that you just inspired a young man to be a soldier?" I told the cadet when I caught up with him.

Of course, the cadet was surprised that a three-star General approached him. But I wanted the cadet to know that he was making a significant difference in a young boy's view of our

Army, just by the way that he carried himself. I told him that even though he hadn't realized it, he was a role model.

The cadet replied, "Thank you for letting me know, Sir."

I took that West Point cadet to meet his young admirer and to take a picture with the boy's "real" Army hero.

The point is that you do not need to be a General, a CEO, or someone with many years of experience to be a great role model for others. You may already be one.

#7 FIND THE SPECIFIC TALENT YOU NEED

A leader must personally engage to find specific future leaders. When I became Chief of Engineers, we only had one other minority General Officer, a female Brigadier General.

I knew that we had to build a pipeline of highly talented women who would rise to the senior positions in the U.S. Army Corps of Engineers.

In the Army, we grow leaders. The pipeline of talent starts with graduates from West Point or through our Reserve Officers' Training Corps (ROTC) programs located at many universities across the country.

While Chief of Engineers and prior to the departure of my male Aide-de-Camp, I was looking for an exceptionally talented female leader to be included among those nominated for an interview to become my Aide-de-Camp.

We started by first identifying the number of women cadets who had selected the Engineer branch. Next, we reviewed the number of women Engineer Officers in the pipeline.

While I wanted to always select the best officer to become my Aide-de-Camp, I also wanted to ensure that talented women were included in the interview process in addition to the male nominees.

Becoming an aide meant the candidates had to have time available in their own career development track to take on the role of aide. So, I identified three different year groups of women that might have the time available in their career development to be an aide. The number of highly competitive women in three different year groups of officers graduating in the mid-1990s was one, zero, and two respectively—in the entire Army.

I asked my current aide to help find his replacement by setting up interviews with both women in the year group that included the two highly competitive women as well as exceptionally talented male officers in those year groups.

But, initially, success eluded me.

Days later, my aide said that one of the officers we had identified was going into a key tactical assignment that was important for her career. I understood this and ruled her out. The other officer replied that she was not interested because she was going into a second tactical level assignment. I could have pushed the issue but elected not to because the Aide-de-Camp job is very demanding and the officer must be willing to serve in the position. So, I selected a well-qualified male officer to become my Aide-de-Camp.

The next year, I tried again to find a highly competitive female Engineer Officer to interview for the Aide-de-Camp position. This time, in the year group that I was considering, there was only one female officer available. My aide arranged an interview. When the Major came into the interview, I thought I recognized her.

"Why do I know you?" I asked. "Have we worked together before?"

She said, "No, we have not worked together."

I asked again, "Then, why do I know you?"

She then sheepishly admitted that I had asked her to interview for the Aide-de-Camp position the previous year, and she had declined so that she could move into a second tactical level job.

As it turned out, I eventually selected this Major as my Aide-de-Camp. I had achieved my objective. I had not only found the female leader who would serve with me as I closed out my tenure as Chief of Engineers, but she would later serve in an assignment with a higher level of rank and responsibility. Following her extraordinarily successful performance as Aide-de-Camp, she was later promoted to Lieutenant Colonel and selected to command an Engineer Battalion in Hawaii. This was a significant accomplishment.

During my fifteen years as a general officer, I had twenty-five aides and numerous Executive Officers. There were many excellent male and female aides. Some have gone on to become Colonels, General Officers, and civilians—all successful. These Aide-de-Camp and Executive Officer positions offer a wonderful opportunity for young and exceptionally talented officers to learn about their profession at the corporate level. However, finding the most talented young leaders, particularly in small population groups, takes a targeted approach, vision, imagination, and often, unwavering perseverence.

WHAT I LEARNED

1. **Start with yourself.** Leverage your passion and leadership style so that you best understand how to assist others.

2. **Find a role model.** Role models can offer a glimpse into the promise the future holds.

3. **Find your passion**. It has been said that if you love your job, it's not really work. However, finding your passion may not be simple. It may develop over time. It requires discovery—knowing yourself and your talents, then matching them to a life's work.

4. **Be persistent.** There is generally always a way to reach your objective. One just needs to find the right path.

5. **Be ready for an unexpected mentor**. A mentor could be someone you don't know, but who has an interest in your success. Be open to offers of guidance, advice, and support.

6. **Become a role model.** One does not need to be older, wiser, and more experienced to serve as a role model.

7. **Find the specific talent you need.** Finding talent is hard work. Finding talent in smaller population groups is even

more challenging. The leader must personally engage in searching for and developing talent.

The single biggest way to impact an organization is to focus on leadership development. There is almost no limit to the potential of an organization that recruits good people, raises them up as leaders, and continually develops them.

~JOHN MAXWELL

CHAPTER 16

CRISIS MANAGEMENT AND LEADERSHIP
HURRICANE SANDY: A CASE STUDY IN CHALLENGES AND SOLUTIONS

By failing to prepare, you are preparing to fail.
~BENJAMIN FRANKLIN

FORESIGHT IN LEADERSHIP

In May of 2012, President Obama held a meeting in the White House Situation Room. Present were the federal government's first responder team, which is a group of men and women at the highest level designated to lead national crisis management and recovery efforts, especially natural disasters such as hurricanes.

On that day, it was a formidable group that gathered around the conference table in the basement of the West Wing to discuss a first-level response plan for hurricanes. Present were the Homeland Security Secretary Janet Napolitano, Homeland Security Advisor John Brennan, Federal Emergency Management Agency Administrator Craig Fugate, Treasury Secretary Jack Lew, Deputy Energy Secretary Dan Poneman, and several others. I was also in the Situation Room that day, representing the U.S. Army Corps of Engineers along with my Director of Contingency Operations, Ms. Karen Durham-Aguilera.

I vividly recall President Obama telling the group in the Situation Room that he was concerned that we might have a significant storm on the northeast coast of the United States, and he wanted us to prepare for it.

Little did we know that just five short months later the United States would be hit by one of the deadliest and most destructive hurricanes in recent history—Hurricane Sandy, or as it came to be known, Superstorm Sandy.

MEGA RESPONSE TO A MEGA CRISIS

Hurricane Sandy made landfall on the northeast coast of the United States on October 29, 2012. We were not prepared for the magnitude of the disaster that Sandy caused. Before it was spent, Superstorm Sandy would impact the entire eastern seaboard of the United States. The hurricane affected twenty-four states, resulted in forty-eight deaths, and caused over $70 billion in damage.

Several of the members of the senior leaders in the Situation Room had been there before. Seven years earlier, in August 2005, the United States had been hit by Hurricane Katrina. This hurricane was responsible for over 1,800 fatalities, left millions homeless, and cost $108 billion in damages, making it one of the deadliest hurricanes to ever reach the United States.

As a result of the Katrina devastation, we learned the importance of planning, preparedness policies, and systems.

To that end, many processes and procedures were already in place that did improve our ability to respond to Hurricane Sandy when it made landfall in 2012.

One of the most important organizational structures put into place by the Department of Homeland Security was the National Response Framework or NRF. The NRF was designed to ensure unity of effort through unified command structures during an emergency, including a natural disaster such as a hurricane.

Under the NRF, clear roles and responsibilities with seamless coordination and communication protocols were established.

This unification took place under the NRF's fifteen Emergency Support Functions (ESF). These functions were designed to carefully outline the organizational structure among the federal agencies.

THE U.S. ARMY CORPS OF ENGINEERS—WARRIOR BUILDERS

Under the NRF's fifteen Emergency Support Functions, the U.S. Army Corps of Engineers was responsible for Emergency

Support Function #3, public works and engineering—specifically electrical power, water, and debris removal.

The U.S. Army Corps of Engineers is uniquely suited to this responsibility. The history of the U.S. Army Corps of Engineers can be traced back to June 16, 1775 when Colonel Richard Gridley became General George Washington's first Chief of Engineers. I was honored to become the nation's 53rd Chief of Engineers. First founded in 1779 to assist in the building of structural facilities for the U.S. Army, the Army Corps of Engineers was later re-established in 1794 by the new federal government and then re-established for the third time in 1802 as a division of the federal government. Historically, the U.S. Army Corps of Engineers proved themselves invaluable in battle, from the Battle of Rappahannock during the Civil War to the Second World War. They played a significant role in the success of D-Day by clearing obstacles from the beaches and building roads for passage off the sands. But it wasn't only in times of war that the U.S. Army Corps of Engineers has delivered. They successfully accomplished countless other challenging projects in time of peace, projects such as designing and building the massive Bonneville Dam; completing the iconic 555-foot tall Washington Monument; constructing The Pentagon, the Library of Congress, and the Lincoln Memorial; working on the restoration of the Great Lakes; and even building the Vehicle Assembly Building that would become the home of NASA's Apollo Space Program.

This was a team suited by history, experience, and skill to play a major role as first responders to a national disaster.

PREPARING TO PREPARE

Leadership at every level is essential during a crisis. To that end, FEMA Administrator Craig Fugate and I hosted several preparation meetings in the FEMA operations center prior to Hurricane Sandy. These sessions included representatives from each of the federal agencies. However, even with the best preparation in place, there are always significant challenges when an emergency leads to devastation that is as all-encompassing as that created by Superstorm Sandy.

CRISIS-GENERATED CHALLENGES AND SOLUTIONS

No matter how careful the preparations are, there are always unexpected challenges, issues, and unforeseen problems that arise, especially during crisis situations. During the time of Superstorm Sandy, I was often reminded of what Martin Luther King, Jr. said: "The ultimate measure of a man is not where he stands in moments of comfort, but where he stands at times of challenge and controversy." We were faced with several "challenges and controversies" during the subsequent attempt to restore order and services in the immediate aftermath of Superstorm Sandy. Some of these challenges were physical, due entirely to the hurricane itself; others were due to the inevitable and often conflicting forces between people. But every single challenge taught a significant leadership lesson.

CHALLENGE #1 INTEGRATING NATIONAL RESPONSE WITH LOCAL NEEDS

In a mega-crisis, it is essential that one leader has the authority to make operational day-to-day decisions.

That's exactly what happened in the Superstorm Sandy emergency.

As soon as Hurricane Sandy made landfall in New Jersey, President Obama along with all the key Cabinet Secretaries, met at the FEMA operations center and connected with local leaders through a Video Teleconference Conference (VTC). We received updates on the situation from New York Governor Cuomo, New Jersey Governor Christie, New York City Mayor Bloomberg, in addition to the mayors of surrounding towns affected by Superstorm Sandy.

At the very beginning of the VTC, I recall President Obama's directive: "If you receive a call from one of the Governors or Mayors for assistance, you have thirty minutes to respond to their call."

The President committed the federal government's proactive stance to the state and city leaders and assured them that they would be provided with the assistance they required during the crisis response.

After the Governors spoke, the President called for any questions from the Mayors. At that moment, it seemed as if every Mayor on that call spoke at once. But the one dominant voice was that of Mayor Zimmer, the Mayor of Hoboken, New Jersey. She was fearless. She told President Obama that she had requested help but no help came. She then pleaded with the President for one important act of assistance. She asked for help to untangle the transportation blockages resulting from Superstorm Sandy so that people could travel across the Hudson River from Hoboken to their jobs and offices in New York. This help would enable people to return to work and some sense of normalcy. President Obama sensed the desperation in her voice and directed the federal government leaders to figure out a way to restore traffic routes into New York City by the following Monday.

That was a tight deadline.

And then that deadline became even tighter.

After the video teleconference ended, President Obama made a startling pronouncement.

"I've changed my mind," he announced. "Rather than the original thirty minutes, you all now have just fifteen minutes to respond to any request that comes from our local leaders." He went on to say, "And if these local leaders feel they have to call me to obtain federal government help to support their request, then you're in trouble."

After the meeting with the President, my first task was to immediately travel to Hoboken to meet with Mayor Zimmer.

CHALLENGE #2 RESTORATION OF NORMALCY

While sometimes disruption is critical to creating an atmosphere that triggers out-of-the-box thinking and innovation, there is a fine line between creativity and chaos.

HOBOKEN, NEW JERSEY

Mayor Zimmer and I visited the railway station at Hoboken. What we saw was a serious problem. Railcars had been flipped over by the storm and were piled on top of each other. It would take some time to reopen this primary artery for travel to and from Manhattan.

That wasn't all. Next, we traveled to the ferry site because ferry boats provided an important link across the Hudson River between Hoboken and Manhattan. The ferry site was also severely damaged, but it was recoverable.

However, before any transportation conduits could be restored, we faced the critical challenge of the lack of electrical power. All power was out.

The restoration of power at the ferry site became our primary objective. And within twenty-four hours, the Prime Power Engineer Battalion of the U.S. Army Corps of Engineers had proven themselves more than up to the task. They restored electrical power. Through the teamwork of local and federal partners, as President Obama directed, commuter traffic to and from New York City, albeit limited, was restored.

Crisis Management and Leadership

BROOKLYN-BATTERY TUNNEL

One of the most significant missions for the U.S. Army Corps of Engineers involved the removal of all the floodwater from the Brooklyn-Battery Tunnel.

The longest continuous underwater vehicular tunnel in North America, the Brooklyn-Battery Tunnel was first opened on May 25, 1950. After Superstorm Sandy, both two-lane tubes, one in each direction, were filled with water due to the storm surge.

Getting this tunnel back into operation to handle vehicle traffic was the highest priority for Governor Cuomo, who visited the site almost daily during the initial response.

The original effort by the U.S. Navy's Supervisor of Salvage and Diving involved draining the tunnel using suction pumps from both ends of the tunnel. That process was slow and exceedingly difficult.

A faster, more effective solution was required.

The bottom of the tunnel was about 150 feet below the surface of the mouth of the East River. The best pumps for the job were high "head" pumps that could pump water at a much higher pressure and elevation, or "head," than suction pumps.

The problem was that even though these types of pumps were desperately needed, we didn't have any available.

Then, one night, there was a knock on my door, and a member of the Coast Guard walked in.

He said, "Sir, I understand you need our help."

"I need a lot of help. How can you assist?" I replied.

The Coast Guard leader simply said, "We have high-head pumps."

I dropped everything.

When I responded, I couldn't have been more enthusiastic. "Have I been waiting for great news like this!" I said.

The Coast Guard put their high-head pumps at the lowest point of the tunnel and started to discharge water 150 feet straight upward out of the tunnel and into the East River. We emptied 485,000 gallons of water from the tunnel in just nine days. That amount of water was the equivalent of filling the Rose Bowl Stadium from top to bottom.

Soon traffic was flowing again through the Brooklyn-Battery Tunnel. Teamwork, leadership, resourcefulness, and the right equipment made all the difference.

CHALLENGE #3 ALLOCATION OF RESOURCES

Resources are a finite commodity. During a crisis, leaders must make tough decisions. Who has a resource priority? When should resources be redistributed? And most important of all, who decides on the allocation of resources? During Superstorm Sandy, one of the most important resources was power. And it was the allocation of that power that first solved a critical problem, and then undid the solution.

THE POWER GRID

Long gas lines were highly visible to all Americans in the aftermath of Superstorm Sandy. Gas stations were out of gas. Gas pumps were dry. The challenge was to restore gas deliveries as quickly as possible.

Kinder Morgan is the largest independent transporter of petroleum products in North America. But Kinder Morgan had a big problem. It had no power. Without power, Kinder Morgan couldn't fuel their trucks. And without their trucks, there would be no gas deliveries. Restoring power to Kinder Morgan was one of the highest priorities for both New York's Governor Cuomo as well as the City of New York.

FEMA gave the U.S. Army Corps of Engineers the mission to provide temporary electrical power to Kinder Morgan.

Our Prime Power Battalion brought their largest generators to Kinder Morgan. These were the only type of generators anywhere in the area that could power up Kinder Morgan. In a short time, the terminal was again operational. Trucks were delivering fuel, and the gas lines slowly started to disappear.

Then the electrical grid came back up in New York, and FEMA directed the U.S. Army Corps of Engineers to remove its generators from Kinder Morgan and transport them to another location. The Corps does not act without mission assignments from FEMA, and the FEMA mission assignments come from the Governor or leaders from the Governor's office. We removed the generators.

Unfortunately, the electrical grid went down again.

The problem of getting gas deliveries resurfaced, only this time there was an added complication.

It all started with an email.

CHALLENGE #4 FULL AND OPEN CRISIS COMMUNICATION

Leaders, juggling many critical issues at one time, rely most often on communication. However, communication challenges often arise during times of crisis and that was true of communications issues that came up during Superstorm Sandy. The remedy? Find the communication glitch. Fix it. Move on.

On the way to a 7 a.m. meeting, my Aide-de-Camp asked me if I had responded to an email from the White House that had been sent to me that morning. I said that I had no emails from the White House. My aide handed me a paper copy of the email that the White House sent. Sure enough. An email had been sent at 4:30 a.m. from New York City to the White House then forwarded to me.

It was laced with expletives from New York blasting me for the removal of generators from the Kinder Morgan Terminal and bluntly demanded, "Get General Bostick the hell out of New York City before he screws up something else."

I determined to stay calm and focused on fixing the problem.

First, I informed FEMA about the removal of generators from Kinder Morgan and the crisis that action had caused.

The FEMA Administrator gave us a mission order to send our generators back to Kinder Morgan. I then flew to the Kinder Morgan Terminal site and spoke to the leaders there. I explained that the Army Corps responds to requests from FEMA that come from the Governor's office. The request from the Governor's office had to come from Kinder Morgan. The U.S. Army Corps of Engineers does not independently decide to add or remove generators, or to choose its own mission assignments. I pointed out that the electrical grid was not yet stable; and just because power was back on, the leaders at Kinder Morgan should not request that the generators be removed unless they believe the grid is stable for the long term.

I then went to New York Governor Cuomo's office to explain what happened at Kinder Morgan. The Governor was not available, but his Chief of Staff met with me. I recall him saying, "General Bostick, in any disaster like this, there is always a victim, a hero, and a villain. Guess which one you are?"

I had managed the communication crisis, but I still had to fix the issue of the undelivered email.

I returned to my office and asked my information technology experts why I did not receive the 4:30 a.m. email. My IT experts explained, "Sir, the Chief of Engineers will at times receive inappropriate email, and we have a filter that removes any email with curse words." I said, "I appreciate that, but we're in a crisis, and there are a lot of people terribly upset. Remove the filters."

With the filters removed, I was in the loop where I needed to be and could avert any additional crisis through more direct lines of open communication.

CHALLENGE #5 CRISIS LEADERSHIP VS. CRISIS PERCEPTION

An old adage, "perception is everything," applies to times of crisis. In times of crisis that saying can become an important underlying theme of crisis management. That was the situation I encountered during the aftermath of and response efforts to Hurricane Sandy.

MAYOR BLOOMBERG, NEW YORK CITY

The next challenge came in New York City. I was invited for lunch by then-Mayor Bloomberg to discuss ways in which the U.S. Army Corps of Engineers could provide support for the response effort to restore services in New York City immediately following Superstorm Sandy.

The challenge was massive and so we were joined by the New York City Deputy Mayor of Operations, my Director of Military Missions, Major General Ken Cox, and my New York District Commander, Colonel Paul Owen.

My first attempt at support was in response to the many power outages in New York City. I offered to supply enough generators to provide power while the main electrical supplies were being repaired. Mayor Bloomberg looked at me, took a sip of his coffee, and said, "General

Bostick, we have plenty of generators. How else can you help us?"

My second attempt at providing support was to help alleviate the massive flooding that had resulted from the storm surge and had made many transportation avenues not only impassable but also inaccessible. I offered to provide pre-positioned pumps to help remove water from the subways and many other locations. Mayor Bloomberg responded, "Well General Bostick, I have also pre-positioned many pumps, so I think we're okay in this area."

My third attempt was to assist with the growing piles of trash that were threatening the movement as well as the health of New Yorkers. I offered help with the removal of debris. Once again Mayor Bloomberg looked at me and said, "General Bostick, do you realize how much trash we move in New York City each day? We can handle the trash."

I was just about to come up with other ways in which my team could provide support when Mayor Bloomberg asked me a question right out of the blue. "Can you tell me about your family?" he asked.

"My family?" I replied, surprised at the pivot in the discussion. "We're in the middle of a crisis. Why do you want to know about my family?"

Mayor Bloomberg responded, "Since we're going to be working with each other, I'd just like to know a little bit about you."

I complied. I told him about my family. He listened intently. And then he turned to his Deputy Chief of Operations and asked him to pick up the tab for the troops.

A quick, "Goodbye, General Bostick," and the Mayor stood up and started to make his way to the door.

As Mayor Bloomberg was leaving, the Mayor's Deputy Chief of Operations looked at me and said quietly, "General Bostick, I need to escort Mayor Bloomberg to his car. I'll be right back. Please wait."

Five minutes later, the Deputy Chief of Operations came running back in, almost out of breath. "General Bostick," he announced, "we need it all. Help with generators, pumps, and with debris."

I was surprised. The Mayor had just turned down my offer for the exact same things—generators, pumps, and trash removal.

What changed?

I didn't have to wait long for my answer.

The Mayor's Deputy Chief of Operations explained. "The Mayor just does not want you running around the city on a horse as if you're in charge. Like that General after Hurricane Katrina." In a flash, I realized that the Deputy Mayor for Operations was referring to Lieutenant General Russel L. Honoré best known for serving as Commander of Joint Task Force Katrina. The task force had been responsible for coordinating military relief across the Gulf Coast area impacted by the hurricane. General Honoré, sometimes known as the "Ragin' Cajun," had gained media celebrity and accolades during the cleanup after Hurricane Katrina for turning around the recovery operations. It was felt his direct, no-nonsense style got things done when compared to what many felt were the empty promises of civilian political leaders. The political leaders did not seem to be in charge from the perspective of many watching the crisis unfold.

I immediately understood the message from Mayor Bloomberg and the Deputy Chief of Operations. It was all about crisis leadership versus crisis perception. They wanted and needed the help, but they did not want me to become the "General Honoré" of New York City. In the end, we had a successful and effective working relationship with Mayor Bloomberg and his team. Mayor Bloomberg invited our team back to NYC for the July 4th celebration, and he thanked the U.S. Army Corps of Engineers for its support during the aftermath of Superstorm Sandy.

CHALLENGE #6 MEDIA MANAGEMENT

During a crisis, the media can be a double-edged sword. On the one hand, the media can play an important role in keeping people informed and safe. On the other hand, the media can cause confusion between media spokepersons and the political lines of authority and responsibility at the local level. A leader must be aware of both the pros and cons of media during a crisis.

This message was further driven home after I was interviewed by Fox News, CBS Evening News, and other media outlets about recovery efforts—particularly the CBS Evening News Report.

I was visiting our North Atlantic Division in New York when I noticed Scott Pelley, the CBS Evening News Anchor, waiting near the Division's headquarters at Ft. Hamilton, New York. As I was about to board my helicopter to survey the flood damage over New York City, I felt a tap on my shoulder. It was Scott Pelley. He asked if he could join me on my helicopter. I thought for a moment and quickly reflected on something that Tom Ricks, the two-time winner of the Pulitzer Prize for National Reporting, had observed—it was the difference between the Army and Marines as related to the media. Ricks said that if an Army leader saw a reporter approaching, he would turn in the opposite direction to avoid contact. The Marine leader, on the other hand, would run up to the reporter and use that opportunity to speak positively about the Marine Corps. With the concept of embedded reporters in Iraq and Afghanistan, Army and Marine leaders were used to having reporters around, but some reporters probably believed the Marine Corps was still more open to the media. So, with this story in mind, and in the spirit of transparency, I invited Scott Pelley to join me on the helicopter. I was pleased that the Scott Pelley interview played on the CBS Evening News. I believed that opportunity helped the Nation understand the challenges we faced with the Superstorm Sandy crisis, and educated the Nation as to the federal government's responsiveness.

I was right, but not completely.

Shortly after the media stories ran, I received a call from the White House. I was informed that Governor Cuomo, the

Governor of New York, did not want me or any members of my team to provide any further interviews.

Even though as the leader of the U.S. Army Corps of Engineers I had accepted media interview requests in order to keep the public informed about recovery and restoration activities during a crisis in the past, that turned out to be an unwelcome move.

The year was an election year. Political leaders needed to take the helm and show their constituents that they were firmly in control. Our help was necessary and welcomed, but we had to remain quietly in the background. In the end, our great country and our military understand that the military supports our political leaders. So, despite the normal practice of FEMA and the U.S. Army Corps of Engineers to engage routinely with the public, the political leaders performed this important task. Governor Cuomo and Mayor Bloomberg handled the media engagements, kept their constituents informed, and provided the leadership that successfully managed the crisis.

CHALLENGE #7 TALENT MANAGEMENT AND CRISIS LEADERSHIP

It might seem counterintuitive to consider that a negative crisis is also a positive opportunity for the training and development of individuals and teams. It is important to always consider the element of talent management in every engagement.

This concept of management was true during the Superstorm Sandy crisis. The storm provided powerful lessons in leadership development and learning.

During a crisis, there are never enough people to manage all the required tasks. Therefore, I was determined to use every opportunity and challenge brought by Superstorm Sandy to train additional leaders.

My first training effort was to send my Director of Military Missions, Major General Ken Cox, to New York City. His assignment was to assist our North Atlantic Division Commander, Brigadier General Kent Savre, in engaging with both New York Governor Cuomo and New York City Mayor Bloomberg.

I had two reasons for taking this step.

First, Major General Ken Cox would provide additional leadership to assist in the around-the-clock meetings.

Second, even though Brigadier General Kent Savre was fully competent and capable, sending Major General Ken Cox into New York City would provide an opportunity for senior leader talent management in a crisis.

Next, I brought in two generals who were not directly involved in the response effort but who I thought would benefit from the crisis experience. They were Major General Mike Wehr, our Lakes and Rivers Division Commander, who was based in Cincinnati, Ohio, and our South Pacific Division Commander, Brigadier General Mark Toy, who served in San Francisco, California. I sent Mike to work with New Jersey Governor Christie. Mark's job was to assist in the management of the logistics supply point, which was the depot where we kept our supply of generators, pumps, and other equipment.

My Deputy, Major General Todd Semonite (now a retired Lieutenant General and who followed me as the Chief of Engineers) and I worked closely with Secretary of Defense Panetta, FEMA, U.S. Army leaders, and other executives to manage the crisis.

It was also important to include younger, talented leaders in these crisis leadership training opportunities.

I reached out to Brigadier General Peter "Duke" DeLuca, Commandant of the U.S. Army Engineer School at Fort Leonard Wood, Missouri, and asked if he could excuse about thirty young officers from their coursework. My plan was to deploy them to New York City and New Jersey to assist in the recovery effort. These young leaders were able to contribute to the success of the mission and learn from the experience—leadership lessons that would certainly help them in the years ahead.

Training opportunities were not limited to the military.

Using the crisis as an opportunity to engage some of our civilian leadership, who would not otherwise be involved, was important as well. As an example, we moved one of our Senior Executive Service leaders, Mr. James Balocki, from Washington, D.C., to the heart of New York City to help manage the urban electrical power issues. Ms. Sheri Moore served superbly as my Executive Officer during the crisis response.

CONCLUSION

Preparation for a crisis is critical. However, it is important to understand that not every contingency resulting from a crisis can be anticipated. A crisis will shine a light on weaknesses in an organization, a plan, or a leader. A crisis also validates strengths. And sometimes a crisis can become the source of powerful actions that might have otherwise been neglected or delayed.

This was the case when I met with Governor Christie late one evening in New Jersey. The hurricane was over, but the devastation was just becoming apparent. Even though it was close to midnight, the Governor and his team were hard at work. Governor Christie was a political leader who really cared about his people and worked tirelessly to support them.

Much later that same night just as dawn was breaking, we walked along the coastline. In some areas, all we could see was destruction—homes and buildings destroyed. But just a few miles away homes were intact.

"Why are some areas destroyed and others are not?" asked Governor Christie.

I explained, "Governor Christie, in some areas, the projects on the coastline are authorized by Congress, funding has been appropriated by Congress, and construction could go forward. In other areas, the projects may be authorized, but Congress has not appropriated money for the construction to begin." I went on to say, "After Katrina, many of the projects that had been authorized were finally appropriated. Funds were released. Construction was started and completed."

I recall Governor Christie say, "I want my Katrina money."

Governor Christie received the necessary funds, and the Corps began construction.

It is the combination of pre-planning before a critical event and crisis management during a critical event that results in leadership outcomes that best serve the organization and its people.

WHAT I LEARNED

1. **Visible and engaged leadership is essential during a crisis.** President Obama's message of responding in fifteen minutes was known by all, and the President repeated this point often. The Governors and Mayors repeated this "response time" expectation as well. Brief, key leader messages can stick and make a difference.

2. **Understand the politics of any situation.** It is often better to work behind the scenes and let the political leaders be the face of the response effort.

3. **Media.** The media can be very helpful in carrying the message of any organization. They must be engaged with the facts, assumptions, and the strategy regularly and often, so they can keep the public informed.

4. **Teamwork, leadership, and the right equipment make all the difference.** Each member of the team brings different skills and resources to any successful mission.

5. **Communication.** During a crisis, clear, crisp, consistent, easily understandable, honest, and timely communications are essential.

6. **Education.** In any crisis it is important to educate self, educate team, and educate stakeholders.

7. **Strengths and weaknesses.** A crisis will shine a light on weaknesses in an organization, a plan, or a leader. A crisis also validates strengths.

8. **Leadership matters more than ever in a crisis.** A leader must have both a macro and micro view of the situation. A leader must be adept at situational triage, ensuring the restoration of services in a logical and effective manner that utilizes every available resource.

Faced with crisis, the man of character falls back on himself.
He imposes his own stamp of action, takes
responsibility for it, makes it his own.
~CHARLES DE GAULLE

CHAPTER 17

LEADERSHIP AND PEOPLE— A WINNING COMBINATION

Leadership is not about being in charge. Leadership is about taking care of those in your charge.

~SIMON SINEK

MILITARY STRATEGY VS. BUSINESS STRATEGY

There are many differences between the world of war and business. An article in the *Harvard Business Review* observes that: "Despite the oft-cited analogy between warfare and business, military principles clearly can't be applied wholesale in a business environment. The marketplace is not, after all, a battlefield . . ."

And yet, there are also many similarities.

Forbes Magazine reports that "it is difficult to dispute the (at least) structural similarities between the two venues: strategy, logistics, resource allocation, competencies, recruitment and retention, intelligence, leadership, and the list goes on." The article goes on to quote retired Lieutenant Colonel (U.S. Army) John Nagl who said, "I think there are similarities. Leadership really matters. A focus on the objective is necessary. Both are competitive exercises in which you operate under conditions of uncertainty. Innovation matters, and the ability to see around corners and predict, to at least a certain extent, the shape of the future, is imperative. All these things in all of these ways, I think, are similarities and parallels between business and warfighting."

War and business . . . similar or different? Which is it?

The fact remains—they are both.

And as such, there are valuable lessons from the battlefield that would be helpful in the board rooms and leadership suites of business.

MISSION FIRST, PEOPLE ALWAYS

There is a phrase we use in the military: "Mission First, People Always." Simply put, the mission always comes first, and that is true whether in war or business. As General MacArthur clearly stated in his famous "Duty, Honor, Country" speech, the mission of the Army is to win our nation's wars.

To achieve this goal and fulfill the mission, leaders must always take care of their people. Similarly, whether on the battlefield or in business, people must feel secure with their leadership.

When men and women go into battle, they go knowing their lives are at risk. A leader's job is to ensure that soldiers are well trained, well equipped, and well led. That job is what military leaders mean by taking care of soldiers. If the troops know that leaders have their back and genuinely care for them, those soldiers will go to the ends of the earth for that leader. On the other hand, leaders who are focused only on themselves, on winning at all costs, even at the cost of their troops, will ultimately lose the faith and confidence of their soldiers, which puts the mission at risk.

How soldiers are treated makes a difference.

During my first year at West Point, we had to memorize Schofield's definition of discipline. Lieutenant General John M. Schofield was an American soldier who led his troops during the American Civil War. He later became U.S. Secretary of War and Commanding General of the U.S. Army. Lieutenant General Schofield wrote his definition in 1879, and it has always been one of my favorites. It is a powerful statement of the relationship between a leader and the people who serve: "The discipline which makes the soldiers of a free country more reliable in battle is not to be gained by harsh or tyrannical treatment. On the contrary, such treatment is far more likely to destroy than to make an army. It is possible to impart instructions and to give commands in such a manner and such a tone of voice to inspire in the soldier no feeling but an intense desire to obey, while the opposite manner and tone of voice cannot fail to excite strong resentment and a desire to disobey."

Professionally and personally, I have followed that definition of leadership closely throughout my career and as a result have built tight, motivated, and extraordinarily successful teams.

A DIFFERENT KIND OF GENERAL THAN EXPECTED

When I was the Chief Operating Officer in a publicly traded company, a colleague once said to me, "You're a different kind of General than we were expecting."

"How so?" I asked.

My colleague replied, "We thought you would come in barking orders, yelling, and getting mad at people, but you're very calm, and you do not raise your voice."

I said, "There are all kinds of Generals in the Army with every type of personality and approach toward leadership. My style has worked for me." Then I went on to say, "On our first day in Iraq, we had eight soldiers killed and fifty-two wounded. So, it is easier to be calm when faced with the day-to-day challenges of our company. No one is shooting at us."

REPLACE "SOLDIER" WITH "EMPLOYEE"—REPLACE "BATTLE" WITH "BUSINESS"

Going back to Schofield's definition of discipline, a business leader may consider replacing "soldier" with "employee" and "battle" with "business." How would the other leaders in your business respond to you? How do the employees in your business respond to you?

There is a difference between leaders who are demanding yet inspiring and those who are harsh and tyrannical in pressing their teams to achieve success. All employees want to be on a winning team, and they want to succeed. They want to see increased revenues and excellent quarterly earnings. They will often work harder for the right type of leader—one who treats them with dignity and respect rather than a leader who has a harsh and tyrannical style of leadership.

What is the right balance in business?

Perhaps the right balance is, "Shareholder interest first, people always;" "Customer interest first, people always;" or stay with "Mission First, People Always." CEOs and Board members must decide. As a leader, you might ask yourself some tough questions: Is your organization always thinking about its people? Are you managing talent through training and leader development? Are you building diverse teams that include many varied and different types of employees who can also contribute to the mission of the organization? Is your team, your leadership, and your Board representative of America? Do the people on your team feel safe enough to learn from their failures? Do they feel comfortable speaking up? Individuals and teams must learn from their challenges so they can enjoy the experience of winning after losing.

LEADERSHIP ROLE IN TALENT MANAGEMENT

There is a close relationship between a senior leader and the junior leaders who work for him or her.

These junior leaders have a responsibility to assist the senior leader in executing the goals of the organization successfully. But the senior leader also has a responsibility—to mentor and train these junior leaders so they can themselves become leaders at a higher level.

Chris Hadfield, an astronaut and a former Commander of the International Space Station described leadership: "Ultimately, leadership is not about glorious crowning acts. It's about keeping your team focused on a goal and motivated to do their best to achieve it, especially when the stakes are high, and the consequences really matter. It is about laying the groundwork for others' success, and then standing back and letting them shine."

Talent management is key in selecting junior leaders with the potential to serve at more demanding levels of leadership. However, for me, choosing a junior leader as my Aide-de-Camp equated with my experience in purchasing a new car. I did not like the process involved in either experience.

When I was serving in Iraq, I had four aides in just over a year. Most Generals might have one for the entire deployment.

Having just one aide would have been optimal for me—but not necessarily for my aides or for the organization. Why? Because I strongly believed that my aides were not only there to support me but to grow in their own leadership abilities.

It was my job not only to be supported by my aides, but also identify their leadership abilities, help them hone their leadership skills, and set them free to become stronger leaders. And they did.

Starr Corbin, Mike Rainey, Paul Hicks, and Frank Myers were my aides in Iraq, and they were all excellent.

Starr was one of my longest-serving aides during our deployment to Iraq. She served with me for eight months, four while we were in the 1st Cavalry Division and another four months during my time as the senior engineer in Iraq. She was a signal officer, but as excellent an aide as she was, she had to go back to serving as a signal officer in the 1st Cavalry Division, so I let her return to the First Team.

Then Mike Rainey became my aide. Mike was a combat engineer and a real warrior. During my counseling sessions, I always asked my aides about their future. Mike told me he wanted to teach in the Systems Engineering Department at West Point. Having taught engineering at West Point, I knew this opportunity would be great for Mike, and he would be a superb role model for the cadets. To prepare for his dream post, Mike planned to finish his deployment with me in the summer of 2005 and then go to graduate school in the Fall of that same year. I called West Point to find out what they were planning for Mike. The Systems Engineering Department planned on Mike going to graduate school in January of 2005 so that he could be prepared to teach cadets in the fall of 2006. Even though he was an excellent aide, I released Mike after three months to attend graduate school.

My next aide was Captain Paul Hicks. I had called Human Resources Command and asked for the best Engineer Captain they could recommend for the position as my aide. They recommended Paul Hicks. Paul had just redeployed from Afghanistan three months before and was currently serving with the U.S. Army Corps of Engineers in Texas. I was in Iraq at the time. I called Paul and asked if his wife would be okay with him deploying to support me in Iraq. Taking family needs into account is an important element of talent management, particularly for a demanding assignment apart from the family. Rarely does talent operate in a vacuum. Paul checked and his wife was supportive of his deployment into Iraq. In addition to serving as a superb aide, Paul was a gifted guitar player. I also played the guitar, but I was not nearly as talented as Paul. Still, we practiced together, and I learned from him. So, the aide was not the only one learning from the experience—it worked both ways. Then one day Paul came into my office and asked if he could go home temporarily for the birth of his second child, which he estimated was due in about three weeks. He had missed the birth of his first child when he was serving in Afghanistan. I said, "Absolutely!" Paul was all packed up and ready to leave when he came into my office again, and this time said, "Sir, no need for me to go home. My wife had our baby this morning." I looked at him, then said, "Okay, you've served your country in Afghanistan and Iraq. Pack your bags, redeploy home, and help your wife with that new baby." In the Army, I've always believed that "we enlist a soldier and retain a family." Paul was one of our best, and we had to retain him by supporting him and his family. Paul had served as my aide for three months.

Following Paul Hicks, Frank Myers then served as my aide. He was a Reserve Component Officer with a law degree and a master's degree in business administration. Frank served through the rest of the deployment for about three months. His days were just like my days. Each day in Iraq started with physical training at 4:30 a.m. in the gym. This was followed by a workday that rarely ended before midnight. That was our routine, seven days a week, although we would sleep in and exercise later in the day on Sunday. Aides always arrived at work before the General and remained in the office after the General to close everything out. I often felt that three to four months as an aide in combat might be the equivalent of a year in peacetime.

Overall, I had twenty-five aides in my fifteen years as a General Officer. I considered the assignments of aides, Executive Officers, and chiefs of staff as an opportunity for developing the future talent of our Army. While still early in their careers, aides and others close to me witnessed first-hand day-to-day operations, different styles of leadership, and gained a broader understanding of the Army. This knowledge made them better informed, more capable leaders when operating at tactical levels and later as senior leaders.

THE BIG PICTURE

In the military, I have found that leaders working in the field and away from higher headquarters are often not aware of the challenges faced by the organization at the most senior levels. They cannot see the big picture. The same is true in business. Whether away from higher headquarters or not, leaders can be myopic in how they think, operate, and approach business decisions based on their level of experience.

A large part of the challenge is how they themselves were trained, especially if they had never served at the corporate level.

Military aides, on the other hand, often serve at the corporate levels of the military. They learn some of the nuances of serving in The Pentagon, engaging with Congress and the media. These experiences give them a unique perspective regarding leadership and prepare them to serve not only as effective tactical leaders but also as outstanding strategic leaders as they rise through the organization.

How civilian leaders are grown, nurtured, and managed can impact their future success. Even if junior leaders are not "assigned" to corporate headquarters, there are methods by which they can gain corporate level experience. I have reached out to junior leaders in the corporate world to assist with developing parts of the corporate budget, preparing for a Board meeting or a quarterly earnings call, and assisting with corporate restructuring efforts. A short, intense engagement opportunity at the corporate level can give leaders experience early in their career that will help them understand the big picture and learn the important skill of asking good questions.

ROCKETS INTO THE PALACE AT THE GREEN ZONE

There are dangers in war. There are risks. The responsibility of a leader is to try to mitigate risk as much as possible, and then to help his or her people no matter what.

This responsibility to mitigate risk and help people can save lives. While in Iraq another soldier and I both learned a valuable lesson in leadership. My office was in the Green Zone, the most common name given to the International Zone of Baghdad where the international community headquarters were located. I had just left a meeting with some of our civilians in the Project and Contracting Office which was in a palace in the Green Zone, and I was walking to my vehicle. Two of the civilians stayed behind to review a project when, without warning, enemy rockets bombed the palace. The rockets struck the two civilians and killed them instantly. I was just twenty-six paces away from the two fatalities. My security detachment grabbed me and started pulling me out of the danger area. I stopped my security detail and redirected them to help me with the casualties. I picked up an officer who was bleeding profusely. With the help of my security detail, we took him to our vehicle and escorted the wounded officer to the hospital. I never knew who that officer was. But at an event years later, an officer approached me and said, "Sir, you probably do not remember me, but you put me in your vehicle when the palace was attacked. I just wanted to finally thank you for never leaving a fallen soldier behind."

RECOGNIZE THE ENTIRE TEAM

As the Chief of Engineers, I often spoke about our great civilians who volunteered to deploy and serve side-by-side with our troops.

From the U.S. Army Corps of Engineers alone, over ten thousand civilians have deployed into combat. These are civilians who, much like our soldiers, have families, friends, and loved ones they leave behind.

These civilians serve in harm's way. A convoy that included several of our civilians serving in Mosul was attacked. Other civilians were injured when a bomb exploded in the shopping plaza in the Green Zone. Many of our civilians were shell shocked when rockets bombed our sleeping area. Our Public Affairs Officer became a medic on the spot and was rendering first aide to the injured. He demonstrated the skills that made him the successful CEO that he is today. One of my civilian contractor security guards was killed. Civilians faced similar dangers as soldiers, but they are often not recognized. I saw the sacrifices of our civilians first-hand–the civilians who make up an important part of our nation's deployed team—and I always thought it was essential to recognize their remarkable efforts.

In business, there are many members of the team who may not be on the "front line" but who are key to the success of the organization. These individuals could be specialists in human resources, legal, logistics, information technology, security, government affairs, public relations, and in many other areas. They are an important part of the organization, and it is essential that business leaders recognize these key specialists along with the rest of the team.

LEADERSHIP IS MAKING THE RIGHT RESPONSE IN THE TOUGH TIMES
Leaders in business will face difficult times. The stock price may drop. Revenues may be disappointing. The company may be missing its quarterly earning goals. How business leaders respond during the most difficult times will send a strong message to the team. Bad things may happen to members of the team. Military leaders have learned that they must be at the most critical point on the battlefield. Similarly, business leaders must be at the most critical location for the company. That could mean driving, flying, or calling the troops, but more importantly "being there," providing caring and compassionate leadership. As leadership authority Peter F. Drucker pointed out, "Management is doing things right; leadership is doing the right things."

WHAT I LEARNED

1. **Schofield's definition of discipline**. Harsh and tyrannical treatment can destroy both an Army and a business if tolerated and allowed to persist.

2. **Talent management.** Talent management is important for the growth and success of the individual, the Army, and for business.

3. **Growing leaders.** Giving tactical leaders strategic experiences is essential for their growth and development.

4. **The Warrior Ethos in the Army can apply to business.** Never leave a fallen comrade behind in war or business.

5. **Leaders must be visible and present.** Leaders must find a way to "be there" during challenging times and make their support and commitment known to their organization.

6. **We're all in the fight. No matter the duties and responsibilities.** Recognize the entire team.

> *Effective leaders are made, not born. They learn*
> *from trial and error, and from experience.*
> ~COLIN POWELL

CHAPTER 18

PUBLIC-PRIVATE
PARTNERSHIPS

*One of the main lessons I have learned during my five
years as Secretary-General is that broad partnerships are
the key to solving broad challenges. When governments,
the United Nations, businesses, philanthropies, and civil
society work hand-in-hand, we can achieve great things.*

~BAN KI-MOON,

FORMER SECRETARY-GENERAL OF THE UNITED NATIONS

THE CHALLENGE OF PROJECT FUNDING

I had the opportunity to meet with Ban Ki-moon to discuss
the challenges of water. We discussed both the dangerous
side of water with the damages brought on by hurricanes and

floods, as well as the positive side given the importance of clean drinking water. Many of the projects of the U.S. Army Corps of Engineers are associated with our water resources projects. One of the greatest challenges to any project is funding. This is a story of just such a funding challenge and how it was overcome through a unique and unexpected partnership.

TOO SLOW AND TOO EXPENSIVE

During my career, I have testified before Congress thirty-three times. It was during one of those congressional meetings that a member of Congress asked me why the U.S. Army Corps of Engineers, which I led, was so slow in completing projects.

This wasn't an unexpected question, and the answer was simple. I explained that a big part of the challenge was the inefficient method of funding Corps projects. On any given day, the Corps has over three thousand ongoing projects throughout the country. The money to complete these projects is uncertain from year-to-year. Congress could choose to fund or not fund a project each year as it reviews the nation's priorities.

The Congressman went on to ask, "General Bostick, how much money would you need to finish the projects where you have ongoing construction?"

I had the approximate figures. "Sir, we would require approximately twenty-three and a half-billion dollars," I replied.

"How much did you receive in construction dollars this year from Congress?" was the next question.

My answer? "We received approximately one and a half-billion dollars. That's a shortfall of twenty-two billion dollars."

I went on to explain, "If you do the math, our projects take an average of over fifteen to sixteen years to complete. So, it is true that we are slow. These projects could be completed faster. However, the inefficient method of funding which, combined with it coming to the U.S. Army Corps of Engineers irregularly and in varying amounts, are a big part of the challenge."

I used the example of building a house. "Congressman, imagine that I wanted to pay for the construction of a house. I would hire a contractor, but because my funds were not all available, I would have paid him to work only on the first floor—not on the entire first floor, but only on the kitchen because I had only enough money for this part of the project. Then the following year, after a twelve-month delay, I would fund the construction of the living room. Six months later I might have the budget to finish the downstairs powder room. A year and a half later I might have the money to rough out the second floor and frame out the bedrooms. Two years later I might put on the roof. You can see how such inefficient and inconsistent funding could delay a project by years. Additionally, the project would cost much more due to mobilizing and demobilizing costs. Also, inflation from year-to-year would increase the cost of goods such as lumber and other materials. Well, that's what is happening to all of our projects."

Even though I explained the root causes of the delays in our projects, I knew that the Congress could not afford to provide the U.S. Army Corps of Engineers with twenty-three and a half-billion dollars.

We needed a different solution. It was time to become more creative.

DON'T JUST THINK OUTSIDE THE BOX—REIMAGINE WHAT IS POSSIBLE!

After that testimony before the Congressional committee, I returned to my office and called a meeting of some of our senior leaders—my executive leadership team—and shared my thoughts.

"It is clear that we will never receive the type of funding from Congress that we need to complete these projects in a reasonable amount of time. We need to think creatively about another approach. Let's review the possibility of bringing the public and private sectors together on some of our projects. We should start with a flood risk management project."

This approach was radical, particularly for a flood risk management project.

A public-private partnership, or P3, had never been attempted before for a flood risk management project. Flood risk management is one of the top priorities of the U.S. Army Corps of Engineers. We were breaking new ground. We had no map or guidebook for a P3 with these types of projects. We were going to write the rule book as we went along.

Why start that first flood risk management public-partnership venture?

First, the U.S. Army Corps of Engineers has a long history in the care and management of the nation's water resources. As far back as 1824, the General Survey Act authorized the U.S. Army Corps of Engineers to perform surveys of both roads and canals. A few months later, with the passage of the Act to Improve the Navigation of the Ohio and Mississippi Rivers, Congress assigned the responsibility to the U.S. Army Corps of Engineers. By 1914 the U.S. Army Corps of Engineers had completed the vital Panama Canal project. The U.S. Army Corps of Engineers took on even more responsibilities with the Great Mississippi Flood of 1927. Then, in 1936, the Flood Control Act effectively made the U.S. Army Corps of Engineers the leading federal agency when it came to flood control.

Second, the U.S. Army Corps of Engineers had the research capability. In 1969 the control, management, and protection of America's water resources were expanded to include an all-important research component. It was in that year that the U.S. Army Corps of Engineers, with the approval of the House and Senate Appropriations Committees and the Subcommittees on Public Works, formed the Institute for Water Resources. The Institute for Water Resources was formed in order "to enhance the capability of the Corps of Engineers to develop and manage the Nation's water resources, within the scope of the Corps' responsibilities, by developing essential improvements in planning to be responsive to the changing concerns of our society."

Third, flood risk management was one of the top priorities of the U.S. Army Corps of Engineers. My experience taught me that top priorities get top attention and top resources.

FOCUS IS FUNDAMENTAL TO SUCCESS

Focus is fundamental to the achievement of any goal. I learned long ago that if something is a priority, then someone in the organization must wake up each day with that priority as their primary duty. Our goal was now to create our first P3.

Our first step to achieve that goal was to realign duties so that we had two members of our team focused exclusively on public-private partnerships.

Our next step was to find experts. My philosophy of leadership has always included the belief that it is important to learn from the best. I remembered how we had reached out to Master Sergeant Patrick Daize to assist with the almost impossible ninety-day goal of constructing a massive airfield in Tuzla, Bosnia. Patrick's expertise had made all the difference with that project.

I needed to find the same kind of expert for this new venture. And I found just such an expert in my good friend, Fred Meurer, the City Manager of Monterey, California. Fred had been the key to the first-ever public-private partnership with the Department of Defense for family housing on military installations. I was confident that his expertise and experience would help us reach our goal quickly and effectively.

Fred's expertise had grown as the result of a tragedy.

THE LEGACY OF DANNY HOLLEY

I was a graduate student at Stanford University earning master's degrees in both mechanical and civil engineering when I learned of the tragic death of Danny Holley, a thirteen-year-old boy who lived near my hometown of Monterey, California. Danny's father was a soldier serving in Korea. His family moved to Fort Ord, California, to wait for him to return to the United States. Unfortunately, there was no government housing available for the Holley family on the military base at Fort Ord, so the family had to find a place to rent off-base.

Fort Ord is located in Monterey County, California, which is one of the most expensive real estate markets in the entire United States. It boasted residents such as Clint Eastwood in nearby Carmel. Celebrities and power brokers have regularly flown in to play golf at the exclusive Pebble Beach Golf Course. It is a beautiful Pacific coastal area but prohibitively expensive.

Danny's family struggled to make ends meet. Times became more and more challenging for this soldier's family, and the pennies Danny managed to earn collecting and selling cans just weren't enough to feed his mother, sister, and two younger brothers.

So finally, hungry and in despair, Danny left a suicide note saying, "If there was one less mouth to feed, things would be better." And then he hanged himself. Something had to be done immediately to address this tragedy and the housing and financial hardship faced by the Holley family as well as other military families.

It was Danny's death that mobilized both the Army and the local community to recognize the poverty and housing crisis suffered by the families of soldiers serving in our military.

The Army had to make haste and build additional permanent housing on the base. And the only way to accomplish this goal would be to partner with the private sector.

Coordinating this first-of-its kind public-private partnership was Fred Meurer, who at the time was the public works and housing director at Fort Ord. Fred Meurer had decided to go with manufactured housing because he thought it would be very fast; and if it did not work, they could clear them rapidly. Politically, the manufactured housing interests put a lot of pressure on the Department of Defense (DOD) to develop an approach to deal with the DOD-wide housing crisis. This first public-private venture was accomplished with "lease authority," which allowed the leasee to provide the military construction of homes in lieu of receiving cash rent for the land leased by the leasee. Those developing the Army land for immediate housing received the land for free. Fred leased more than fifty acres for 225 units. The time period from award of the lease of the land to occupancy of the homes was about six months. These homes have been 100 percent occupied since they opened on August 30th, 1985. Even though Fort Ord closed in 1994, the mobile home community is considered a quality place to live on the Peninsula.

Fred is a visionary leader. He took on the mission of addressing the severe housing shortage at Fort Ord by bringing the military and private sector together.

Not only was Fred a visionary, but he is one of the most creative leaders I have ever met. Thus, the first public-private partnership related to military housing was born.

THE MONTEREY MODEL

Fred's creative and revolutionary concept is considered the forerunner of the Residential Communities Initiative (RCI) which resulted in the high-quality housing located at all military installations today. RCI became a model for how a public-private partnership can work and achieve a successful outcome for both sides. The authorization for RCI was passed by Congress as the Military Housing Privatization Act in 1996. Fred Meurer once told me that this law should be called the Danny Holley Act.

Later, as the Public Works Director for Monterey, Fred obtained Congressional approval for the City of Monterey to provide the local military base with facility maintenance as a pilot public-to-public project. That is one public entity, Monterey Public Works, partnering with another public entity, the military. This initiative was also a great success. Estimates indicated that it reduced installation costs by twenty-three percent since the project was launched in the late 1990s.

Today, many Department of Defense, public communities, and private sector companies refer to Fred Meurer's work as the "Monterey Model."

THE CHALLENGE OF PUBLIC-PRIVATE PARTNERSHIP FOR FLOOD RISK MANAGEMENT

With this level of accomplishment to Fred Meurer's credit, as the Chief of Engineers I knew the expert I had to reach out to for assistance with public-private partnerships was Fred Meurer.

Fred visited our headquarters and met with our leaders. We learned about the many key factors in the success of the "Monterey Model" and his work at Fort Ord. Most of all, we learned that someone at the local level had to make the public-private partnership work. We needed our own "Fred Meurer" who would be the passionate advocate on the ground for advancing a change in how we funded projects through a P3.

That was our first challenge. With most national-level programs, passionate local leaders must demonstrate success in a way that can be scalable across the country.

Our second challenge was funding. Public-private partnership for flood risk management projects were difficult to monetize. For Meurer's housing project, the funding stream was the soldier's basic allowance for housing, which reduced the financial risk for private partners who committed the upfront cash requirements. For Meurer's facility maintenance project, the military installation was going to pay for these expenses anyway; but through the partnership with the city of Monterey, Fred was able to save overall costs for the installation and for Monterey by ensuring a greater quantity of service. This partnership also helped to bring the community of Monterey closer together with the leaders and family members of the military.

However, with flood risk management projects these elements were not so evident. It was difficult to monetize a levee, for example.

The solution? We launched an outreach program.

The U.S. Army Corps of Engineers reached out to our local divisions asking for nominations for potential public-private partnership projects. One project stood out—the Fargo-Moorhead flood diversion project.

THE FARGO-MOORHEAD FLOOD DIVERSION PROJECT

The communities of Fargo, North Dakota, and Moorhead, Minnesota, battle flooding every year. Unfortunately, any flood risk management project required a benefit-cost ratio high enough to receive federal funding. The flood risk management project in Fargo-Morehead never qualified for federal funding based on the benefit-cost ratio method of prioritization of projects. So that's when the two communities came together to develop a public-private partnership that would serve as a pilot project and satisfy both the U.S. Army Corps of Engineers and the federal government.

The Fargo-Moorhead flood risk management project called for an embankment or dam to temporarily store floodwaters as well as a diversion channel to divert floodwater around both cities.

One of the key elements of success was the fact that the people of Fargo-Moorhead themselves were ready for a change. The local community was all in. They were tired of fighting to save their homes and communities alone every time the Fargo-Moorhead area flooded. The people supported the concept of a public-private partnership. The funding stream for the public payments would come from local taxes to repay investors who were eager to invest dollars into this worthy cause.

To expedite project delivery, the Corps and local sponsors developed a "split delivery" approach. The local sponsors would use a "design-build-finance-operate-maintain" public-private partnership structure to create the diversion channel, and the U.S. Army Corps of Engineers would use its traditional methods to construct the southern embankment or dam. It was a clean, clear, and elegant strategy.

We invited the local Army Corps team, as well as the local community leaders and investors to Washington, D.C., to discuss their plan. I worked with Senator Hoeven (North Dakota), and we met with Mr. Ali Zaidi from the Office of Management and Budget. Senator John Hoeven, Senator Heidi Heitkamp (North Dakota), North Dakota Governor Jack Dalrymple, Congressman Collin Peterson (Moorhead, Minnesota), other leaders, and I met in Fargo-Moorhead with the U.S. Army Corps of Engineers local team, the investors, and the local citizens to work on and achieve an agreement.

Thus, the first and only flood risk management public-private partnership was created.

As this plan came together, I was reminded of the quote from author Ruskin Bond who observed how "In times of crisis, it's wonderful what the imagination will do." While the Fargo-Moorhead public-private project is still underway, it demonstrates that a P3 with a flood risk management project is possible.

As I reflected on this first flood risk management public-private partnership, it was clearly another example of great teamwork, passionate and innovative leaders at every level, a commitment to change, and a willingness to take the risk necessary to create a better future—to reimagine. Change is never easy, but times of crisis offer an opportunity to make significant changes in the way organizations do business and hopefully, it will not require the loss of life, a story like that of Danny Holley, before leaders recognize and make the changes.

WHAT I LEARNED

1. **Priority**. If you have a strategic priority in your organization, someone on the team must start each morning knowing that that priority is their primary duty and responsibility.
2. **Experts**. Reach out to the subject matter experts who have experience related to your challenge and learn from their expertise.

3. **Crisis**. At times of crisis leaders should act quickly, decisively, and make the significant and, at times, revolutionary, changes required. Demonstrate the capacity to recover from a downturn; be resilient.

4. **National programs.** Highly passionate and successful local leaders can implement programs that serve as an exemplar and later scale across the Nation.

Partnership is not a posture but a process—a
continuous process that grows stronger each year
as we devote ourselves to common tasks.
~JOHN F. KENNEDY

CHAPTER 19

TALENT MANAGEMENT

I am convinced that nothing we do is more important
than hiring and developing people. At the end of
the day, you bet on people, not on strategies.
~LAWRENCE BOSSIDY, GE

M y experience in recruiting and personnel, combined with the need to lead winning teams in peace and war, has given me a much better appreciation for talent management. I believe that my personal experience with how my own talent was managed might be helpful to leaders in business and the military. I was fortunate to have many wonderful mentors who guided me throughout my career. At times, their support was based on our discussions and included the sharing of their own experiences and offering advice. On other occasions, there were those mentors who supported me directly. Both types of mentoring made a difference in my career.

In my thirty-eight years of military service, I spent fourteen years in the Washington, D.C. area, or the National Capital Region (NCR), where I learned so much about strategy, policy, and leadership in a large organization. Early in my career, I was selected as a White House Fellow and served as a Special Assistant to the Secretary of Veterans Affairs, Edward J. Derwinski. The purpose of the White House Fellows program is "to provide gifted and highly motivated young Americans with some first-hand experience in the process of governing the Nation and a sense of personal involvement in the leadership of society." It was a great honor and privilege to be selected for the White House Fellowship. This fellowship gave me the remarkable opportunity to grow as a leader and learn from multiple engagements with veterans, Congress, and the White House.

During my military career, I testified before Congress thirty-three times, more than most General Officers. I engaged with the media frequently and have appeared on Fox News, CNN, CBS Evening News, Fox Business, and yes, *The Daily Show*.

As the leader of U.S. Army Recruiting Command, I was given the opportunity to assist the Army in our marketing and advertising effort. We worked with the major marketing and advertising agencies, which included Leo Burnett in Chicago and McCann Erickson in New York City. I served as the

Executive Officer for Lieutenant General Arthur Williams, who was the Chief of Engineers, and also served as the Executive Officer for the Chief of Staff of the Army, General Eric K. Shinseki for two years. Then I served as one of the watch officers on duty in the National Military Command Center (NMCC) on the Joint Staff in The Pentagon working closely with the future Commander of U.S. Forces Africa, General Kip Ward, who guided our team during 9/11. On that tragic day, our team engaged with the leadership of the country as we executed emergency response operations.

My career was balanced with assignments outside of the National Capital Region. Most of my assignments were with combat troop units, starting with the 54th Engineer Battalion near the former East German Border to the 1st Armored Division located near the border of Germany and France, from the 1st Infantry Division at Fort Riley, Kansas, to the 1st Cavalry Division at Fort Hood, Texas.

My assignments were all educational, challenging, and broadening for me as a leader. I never requested or sought out specific assignments in the Army. All officers considered for battalion and brigade command were asked to prioritize their choices for command if fortunate to be selected. Beyond these two positions, I was placed in a wide variety of assignments with little or no input from me. Often, I was surprised to be asked to serve in some of these leadership positions given my lack of experience. Other leaders were clearly managing my assignments. I received excellent talent management.

I recall attending the retirement of a senior General in the Army. During the General's remarks, he said that he had never served in Washington, D.C., which drew thunderous applause from the audience. I thought that having not served in Washington, D.C. was not something that we want our junior leaders to aspire to. Without question, most soldiers love being around the troops and where the action is. But those troop units need strategic guidance, policies, processes, and resources that require exceptionally talented leaders at the corporate level and for the Army, at The Pentagon.

The desire not to serve at the corporate headquarters does not apply only to the military. I saw this same desire to avoid service at corporate headquarters during my time as a C-Suite officer in a publicly traded company. If an individual never serves at the corporate level of an organization, it is difficult for that leader serving at the local level to "think" and operate strategically and in the best interest of the overall organization. Tactical leaders and some operational or regional leaders generally see the world from their own perspective. They can become myopic with a primary focus on their own interests and those of their teams. Often these leaders have little understanding of the big picture. They can win the battle at their level while the larger organization potentially could lose the war.

While frustrating to see tactical and operational leaders viewing the world through a narrow perspective, this focus is, in part, the result of ineffective talent management. I would often use a soccer analogy in describing how leaders sometimes operate. Everyone wants to play soccer, but someone must organize the team, schedule events, and secure resources. The former is perceived as "fun" and the latter is something that many tactical and operational leaders prefer to avoid. They want to "play soccer," not serve in Washington, D.C., or wherever the corporate headquarters might be. Working at headquarters is perceived as being "no fun." The fact of the matter is that perhaps serving at the corporate level is not as much fun as the tactical level, but corporate-level service provides experiences that will help develop and grow tactical leaders. Most tactical leaders would rather not serve at the corporate headquarters; but in the military, more than 80 percent of the Colonel Officers and General Officers serve in the non-tactical part of the Army. These non-tactical leaders help to run the Army.

One example of a tactically focused officer with a myopic view comes from a training exercise at the National Training Center. I was a Battalion Commander, and one of the battalions attached to our brigade was a light infantry unit. We were preparing for a night attack through an area called Red Pass. The brigade needed the light infantry battalion to move forward by foot to secure the pass so that the heavy armored units could safely transit through Red Pass while conducting a night passage of lines where one unit moves its vehicles and personnel through another unit which is stationary. A passage of lines, particularly at night, is one of the more challenging tactical maneuvers for military units.

During the Rehearsal of Concept or ROC drill, where units rehearse each phase of the operation, the light infantry Battalion Commander requested helicopter support to take his troops to Red Pass rather than moving by foot. Our Brigade Commander explained that the light infantry battalion had to move into Red Pass covertly on foot so that the main effort of the brigade was not revealed. Thus, the helicopters were not an option. The light infantry Battalion Commander continued to press for the need to have helicopter support. He was not accepting no for an answer. Our Division Commander was listening to the dialogue, and he said to the light infantry Battalion Commander: "Listen, you heard the Brigade Commander. You're not receiving the helicopters. You need to get your battalion moving on the ten-kilometer road march to Red Pass." The Division Commander went on to say to the light infantry Battalion Commander, "If we were going to war, I would leave you behind." The Division Commander stormed out of the area, and the light infantry Battalion Commander continued to press the case for helicopters.

The light infantry battalion had to start their road march immediately after the ROC drill was complete. As the evening approached, a battle suprisingly started just short of Red Pass. Rather than quietly walking to Red Pass and securing the pass for the rest of the brigade, the light infantry battalion mounted their wheeled vehicles and drove to Red Pass. The

Opposing Forces (OPFOR) heard the vehicles, flanked them, and destroyed the entire light infantry battalion. The brigade lost the battle before moving out of the assembly area.

How many leaders have you seen who cannot see or understand the big picture? While in this case the leader was a mid-level officer, the same can happen with more senior officers who do not understand or are unwilling to accept the higher-level mission over their own tactical level success. Senior leaders who have remained at the tactical and operational levels for most of their careers have limited insight into the challenges of the entire organization. These tactical leaders never received the talent management that would have given them the experience at the corporate headquarters. I have seen this same narrow and tactical view in both military and civilian organizations.

Sometimes, talent management must be directed. Recall that most people want to keep playing soccer; they want to stay in the tactical world rather than serve at corporate head-quarters. Again, my own experience may be instructive. My Battalion Commander, Lieutenant Colonel Dan Christman, and Major Bruce Scott encouraged me to apply for a White House Fellowship. At the time, I felt completely unqualified to compete for a White House Fellowship. Their advocacy and persistence helped me to make the decision to apply. I applied, was selected, and served as a Special Assistant to the first Secretary of Veterans Affairs, Edward J. Derwinski. Working at the top levels of our government made all the difference years later. Several of the White House Fellows are among my closest friends. The White House Fellowship is one of the finest programs in America. The program selects young talented leaders with the belief that many of those selected will go on to make a significant impact in their professions and in public service to our Nation.

Working behind the scenes in supporting me was another leader, Lieutenant General Joe Ballard. General Ballard was the former Chief of Engineers and Commanding General of the U.S. Army Corps of Engineers. Army leaders were considering moving me from my Deputy Chief of Engineers position to be the Commanding General of U.S. Army Recruiting Command. Army leaders asked General Ballard, also an African-American, what he thought the perception would be of moving me to U.S. Army Recruiting Command and moving the other two African-American General Officers. General Ballard offered his thoughts and the result was that I was reassigned to U.S. Army Recruiting Command. I learned so much about the challenges of recruiting, retaining, and developing talent. This was one of my most rewarding assignments, and what I learned in U.S. Army Recruiting Command would serve me well for the rest of my career both in the military and as a civilian. Despite the excellent experience I had at U.S. Army Recruiting Command, most officers and enlisted soldiers do everything they can to avoid serving in a recruiting assignment. Recruiting Command is avoided by soldiers much like The Pentagon is avoided. Soldiers therefore lose the opportunity to learn from these types of corporate-level broadening assignments.

Managing talent helps the individual and the organization. Talent management also helps teams; teammates learn that their leaders understand that personal growth and development of each person is essential for the continued success of the organization. With proper talent management, where skills are aligned with the proper assignments, it is far more likely that talented individuals will remain with an organization.

WHAT I LEARNED

1. **Experience at the highest level is important for future senior leaders.** Corporate experience can provide tactical, operational, and regional leaders with an understanding of the "big picture," which will help them become more effective leaders at every level.

2. **Mentors play an important role in talent management.** Mentors can make a real and lasting difference in the talent management of young leaders whether as a trusted adviser, trainer, counselor, or cheerleader.

3. **New experiences broaden the talents of future senior leaders.** Experience matters, but often leaders must work in areas where they have no experience in order to gain the experience that will serve them well later in life.

4. **Do not win a battle at the tactical level only to lose the war at the strategic level.** Failure can happen when junior leaders are too narrowly focused on tactical and operational level objectives at the risk of strategic goals of the overall organization.

> *Train people well enough so they can leave, treat*
> *them well enough so they don't want to.*
> ~RICHARD BRANSON, FOUNDER OF VIRGIN GROUP

CHAPTER 20

YOUR PEOPLE ARE YOUR BRAND'S BEST AMBASSADORS

Soldiers are not in the Army. Soldiers are the Army.

~GENERAL CREIGHTON ABRAMS

WHAT DO YOU SEE WHEN YOU SEE A SOLDIER?

This seemingly simple question presented itself on at least two occasions during my career, each with radically different answers. And in both cases, in many ways, part of the answer had to do with branding.

Most people think of brands as icons. They are easily identified. They often appear on every imaginable consumer and organizational product. Brands can be so iconic that words are not needed. Everyone who sees a brand not only recognizes it but can immediately identify the organization as well as the culture, goals, and vision it represents. Brands also evoke a response. Brands are deeply emotional, visceral; when seen, they can bring out the best and the worst in the people.

The Harvard Business Review points out that "forming the foundation of a corporate brand identity is the firm's mission and vision (which engage and inspire its people), culture (which reveals their work ethic and attitudes), and competencies (its distinctive capabilities). These elements are rooted in the organization's values and operational realities."

Understanding the power of branding doesn't only apply to corporations producing sodas, hamburgers, running shoes, or computers. The power of branding applies to military brands as well. And learning how to best market its brand is valuable for any organization—public or private.

The two stories that follow are about the Army brand, the reactions it provoked, and how it was made stronger, more powerful, and able to resonate with virtually everyone who came in contact with it. These stories are lessons in branding.

STORY #1 GREGORY PECK AND THE MOVIE *MACARTHUR*

It was 1976. I was in my second year at West Point when the cadets were afforded the rare opportunity to participate in the making of a movie. The film was *MacArthur*. It was about the corn-cob-pipe-smoking General and Supreme Commander for the Allied Powers in the Southwest Pacific Area during the Second World War. There was great excitement not only because of the movie, but also because the famous actor Gregory Peck, star of *To Kill a Mockingbird* and *The Guns of Navarone*, was going to play MacArthur.

As cadets, we were thrilled to participate in the making of this movie, and we all enjoyed our chance to be part of the action. Some cadets were involved in the filming of the parade scene, while others participated as the audience listening to Gregory Peck deliver MacArthur's famous "Duty, Honor, Country" speech. I was one of the cadets in the audience and couldn't help but be moved and wonder what it must have been like for all those cadets who had come before me—the cadets who were actually in that audience fourteen years earlier on May 12, 1962, listening to General MacArthur himself as he delivered that stirring speech, on what would be his last visit to West Point.

DUTY HONOR COUNTRY—WORDS TO LIVE BY

I learned that making a movie requires many "takes" to achieve the final cut—the film the theater-going audience would see on the screen.

I learned that excellence took some patience. There was one scene at the beginning of the movie in which Gregory Peck had to tell MacArthur's famous joke, which went like this: "As I was leaving the hotel this morning, a doorman asked me, 'Where are you bound for, General?' and when I replied, 'West Point,' the doorman remarked, 'Beautiful place: have you ever been there before?'" At that point, there was supposed to be a lot of laughter in the audience. Why? Because MacArthur had not only graduated first in his class from West Point, but he had returned later in his career to serve as Superintendent of West Point, so he had certainly "been there before." That was the punchline. Gregory Peck patiently went through take after take to get that line of dialogue just right. And we cadets sitting in the audience laughed and cheered and applauded dozens of times before the director was satisfied with our reaction.

MacArthur's moving "Duty, Honor, Country" speech has inspired generations of cadets and military leaders, as well as civilians. In describing Duty, Honor, Country, MacArthur (and then Gregory Peck) said, "Those three hallowed words reverently dictate what you ought to be, what you can be, what you will be. They are your rallying points: to build courage when courage seems to fail; to regain faith when there seems to be little cause for faith; to create hope when hope becomes forlorn."

Another part of the filming that took several takes was the closing scene when Gregory Peck, as General MacArthur, said, "Today marks my final roll call with you, but I want you to know that when I cross the river my last conscious thoughts will be of The Corps, and The Corps, and The Corps. I bid you farewell." Each time Gregory Peck said those words, the cadets gave a standing ovation.

AN UNEXPECTED RESPONSE

The movie *MacArthur* was set to premiere at Radio City Music Hall in New York City, and the Corps of Cadets was invited. We were all extremely excited. I recall taking the bus to New York City with my classmates. Some of us stopped at Mamma Leone's restaurant in Manhattan's Theater District—a destination for some of the best Italian food in the city since its founding in 1906 by Mamma Louisa Leone. We were all sharply dressed in our cadet uniforms. The Mamma Leone's staff treated us with respect, and we were proud to be part of the moment—the great film and the great leader it portrayed.

We marched in formation to the premier, and as we approached Radio City Music Hall, we saw a small group of protestors outside the theatre. They were protesting against war in general and the MacArthur movie in particular. We were met with jeers and catcalls.

I was surprised at the reaction of the protestors. The cadets felt that we were doing something good, something positive in support of the movie. We were proud of that film. We were proud of The Corps. We were proud to be serving our country. We were proud of the uniform we wore. We were disappointed that some people were protesting against the movie. Despite the protests, during most of our march to Radio City Music Hall we were met by welcoming crowds.

A FATHER'S LESSON

After the premiere, I called my father, Master Sergeant Sidney C. Bostick, and told him what had happened. My father never talked about his military experience even though he had fought in Korea and served two tours in Vietnam. When I told my father what happened to us in New York City, that there was a small group of protestors and their anger had not only been against the movie but also seemed to be against us, he asked just one question.

"Were you wearing your uniform?"

"Yes, we were all in uniform," I replied.

It was then that my father shared this lesson with me. "Tom, many Americans did not like the Vietnam War, and when you wear your uniform in public, you're representing a war that they did not support." He went on to say, "If you are in public, you should not wear your uniform."

Though it was difficult to hear this from my father, I understood. Having returned from Vietnam to the United States, he came back to a country where soldiers faced the scorn of many Americans for participating in an unpopular war.

In this case, the Army brand evoked a powerful negative response to a small group of protestors. Little did I know at the time that I would advocate for and achieve a powerful positive response from that same Army brand, our uniform, much later in my career.

STORY #2 THE AIRPLANE RIDE THAT CHANGED EVERYTHING

Fast forward thirty years later. I was leading the U.S. Army Recruiting Command, and we were fighting hard to increase the size of the Army in order to support the wars in Iraq and Afghanistan.

Just as in business, marketing was an important strategic tool for the Army. And our brand—the Army brand—was front and center in our marketing recruitment efforts.

The Army showcased our brand in as many venues as we could. We displayed our brand at NASCAR, at the National Hot Rod Association (NHRA), at the Professional Bull Riding Association (PBR), at the U.S. Army All-American Bowl football game for top High School Athletes, and at many other national and state marketing events.

In spite of all our marketing efforts, I felt that our best marketing could be achieved by our soldiers. I believed that ultimately it was our soldiers and how they carried themselves, how they displayed their pride, and how they were perceived in the thousands of hometowns across America that would make the difference in our recruiting goals.

THE BATTLE DRESS UNIFORM

But there was one big branding problem. Whenever soldiers appeared in public, they had to wear their dress uniform, also called the Class A uniform. And it was that dress uniform that was creating some confusion in the Army brand.

I recall one instance in particular. I was on an airplane, and a passenger mistook me for the pilot. And as if that wasn't enough, the passenger went on to ask me if I could provide her with a pillow. On another flight, a passenger asked me what uniform I was wearing. When I replied that I was wearing the Army uniform, he asked, "Which Army?"

It was clear that many members of the public did not recognize that we were soldiers when we wore our dress uniforms. That dress uniform was not showcasing the Army brand as strongly as it could.

We had to make a change.

It was not just the importance of public awareness of our brand that was important, but that public awareness was vital to achieving our recruiting goals.

We developed a plan.

Instead of dress uniforms, I would make sure my recruiting teams wore their battle dress (BDU) in public, the combat uniform that is easily recognizable by the public.

Our plan had to go through The Pentagon for approval. The request for a travel uniform change would take some time, and that could significantly delay our efforts to make progress in our recruiting mission. So, I assessed the risk of moving forward to implement the plan before final approval had been granted and determined that it was low enough to proceed. Then we took that risk. Our recruiters were to wear their Battle Dress Uniform (BDU) in public immediately, even on airplanes. The Army combat uniform was a powerful brand.

CHANGING A POLICY

After our recruiters started wearing their combat uniforms in public, I received several calls from The Pentagon asking why our soldiers were not following Army policy.

I explained the need for Americans to know their Army. I pointed out that our soldiers are the best marketing tool we have for our Army. It took some time, but eventually the Army changed the travel uniform policy allowing all soldiers, whether they were in U.S. Army Recruiting Command or in other units, to wear the combat uniform in public.

The Air Force, Navy, and Marines soon followed our lead.

THE "THANK YOU FOR YOUR SERVICE" BRAND

Not long after the change in policy, I was on another commercial flight. The plane had just landed. My aide and I, dressed in our Battle Dress Uniforms, were walking through the gate area, when an elderly lady approached me and asked if she could give me a hug. Though surprised by her request, I agreed.

She had tears in her eyes and as she hugged me, she said, "Thank you for your service."

She also gave a hug to my Aide-de-Camp.

Afterward, my young aide asked, "Sir, what was that all about?"

I replied, "She probably recalls a time when soldiers were not always appreciated for their service. She recognized us as soldiers and just wanted to say thank you."

As we walked through the airport, we were stopped several more times by people thanking us for our service; I realized that we had made the right call. The Battle Dress Uniform was a strong, recognizable, iconic brand.

But the public recognition didn't just end at that airport. It was repeated over and over again.

One of the best Super Bowl commercials ever—the Anheuser-Busch Applause commercial—honored members of the U.S. Armed Forces. During that commercial, American soldiers in airports all over the country were greeted with spontaneous applause and expressions of gratitude as they passed their fellow Americans. It was that same powerful response that our troops received with the change of the uniform.

With the change to the Battle Dress Uniform for our recruiters and later for the Army, followed by the other services, our military became more connected with America. And I was reminded of that famous George Washington quote: "A Nation is judged by how well it treats its veterans." Now we could also say that a nation is judged by how well it treats its soldiers.

We had come a long way since my experience at the *MacArthur* premier in 1976. This Army had become America's Army, and the public to this day continues to thank soldiers for their service.

WHAT I LEARNED

1. **Duty, Honor, Country, and the importance of values.** The values of an organization, military or civilian, underpin its foundation and support its success from generation to generation.

2. **Sometimes leaders must make key decisions at the critical time, which may be out of step with normal processes.**

There are times when it is best to make changes the leader knows are right, and then sort out the policy later.

3.   **People in any organization are its best ambassadors.** Whether at junior or senior levels, people within an organization can best market and speak about the company.

> *The secret of change is to focus all of your energy*
> *not on fighting the old, but on building the new.*
> ~DAN MILLMAN

CHAPTER 21

FROM BOSS TO MENTOR TO FRIEND: THE POWER OF RELATIONAL LEADERSHIP

The most important single ingredient in the formula of success is knowing how to get along with people.
~THEODORE ROOSEVELT

THE POWER OF THE PERSONAL

Take this quick quiz. Which leader is the relational leader and which is not?

Scenario #1 Leader One sends an impersonal memo to his team thanking them collectively for a good quarter. Leader Two sends a personal note to every member of the team thanking them for their individual contribution to the achievement of the quarterly goal.

Scenario #2 Leader One calls a meeting in which the goal and how it will be achieved is presented to the team. Leader Two calls a meeting in which the goal is presented to the team, and its achievement is distributed among the individual talents of team members.

Scenario #3 Leader One "checks up" on employees to ensure immediate tasks are completed, deadlines are met, and spending is within budget. Leader Two "checks in" to ensure that the team, as well as its individual members, are doing well, are empowered to do their jobs, and have the resources to achieve success.

Were you able to identify the relational leader in each of these scenarios? If you picked Leader Two you have correctly identified the relational leader.

Even though relational leaders are focused on the goals and mission, they never lose sight of the people who make up their teams. Relational leaders understand the concept of "Mission First, People Always." This concept also means helping people to be the best they can be.

However warm and fuzzy, the bigger question remains. Is this a successful leadership style? *Forbes Magazine* writes that "Relational leadership can be incredibly successful, particularly when it is authentic, empathetic, reinforced through gestures of friendship, and embedded in the culture of a team. It's not the only style of leadership but can be powerful and personally fulfilling."

RELATIONAL LEADERSHIP IN TIMES OF CRISIS

Relational leadership is essential, and its impact and influence are perhaps most apparent during a crisis. This book is being written during one of the worst global crises the world has experienced in over a century. The coronavirus has impacted every aspect of life as we have grown to know it, has forced every culture to adapt to mitigate the deadly effect of the pandemic, devastated economies, upended organizational models, shattered traditional team structures, and reached down into every system from education to healthcare, and to the family we hold dear. With this chaos, leaders ask themselves, what is the best way to lead? What tools should be used to restore some semblance of order? What leadership style should be adopted to keep organizations level and dispersed teams productive? The answer is relational leadership. It is a highly effective and "people always" leadership tool, style, strategy for chaotic and uncertain times.

Why?

Because it is first and foremost about people. Human capital is the most vital resource, and its management is a top priority during times of crisis.

EXAMPLES OF EXTRAORDINARY RELATIONAL LEADERSHIP

I had several mentors who inspired me through their own actions to turn my attention to my people. They proved time and time again by their own actions and the example they set that good leaders focus on the mission, but great leaders focus on the people and the mission.

THE CARING PHONE CALL

I recall one difficult day in U.S. Army Recruiting Command. It was a challenging time as we worked to fulfill the requirement for a surge of troops needed in Iraq and Afghanistan. My Executive Assistant received a call and informed me that the Director of the Army Staff (DAS) was on the phone and wanted to speak with me. Now, the Director of the Army Staff is one of the most senior Generals in the Army. I couldn't understand why he wanted to speak to me. The mystery deepened when I asked my Executive Assistant if the DAS himself was on the phone or if it was his Executive Assistant on hold waiting for me. This was a very important distinction. The Director of the Army Staff was senior to me, and the protocol is that senior leaders should not be placed on hold while waiting for more junior leaders. My Executive Assistant said, "Sir, the Director of the Army Staff is on the phone." My mind was racing with all the possible reasons he could have for calling me directly. The first thought that came to mind was, "What had we done wrong?" The next thought was, "What did he need from me?"

I picked up the phone and received a powerful lesson in relational leadership. I said, "Hi, Sir, how can I help you?" The DAS replied, "Tom, I just wanted to call to see how you are doing." "Excuse me, Sir. Do you want to know how we're doing with achieving our recruiting mission this month?" He said, "No, Tom, I'm calling to see how you personally are doing. You have one of the most challenging jobs in the

Army, and I am just checking in to see how you're doing, and to find out if there is anything that I can do to help." "Well, Sir, I'm doing very well, and we have everything that we need to accomplish our recruiting mission. Thank you."

The call did not last long, but it made an indelible impression and reminded me of an especially important lesson. Here was one of the most senior leaders in the Army calling one of over three hundred Generals in the Army simply to check-in. And he made the call himself. He didn't delegate it to anyone else. He picked up the phone. He made the call. He waited until I was on the line. He demonstrated deep caring for his people and great humility. What a powerful message from a senior leader. This was relational leadership at its basic and at its best.

THE CONGRATULATORY PHONE CALL

After my first year in Recruiting Command, we accomplished our mission. Despite the extraordinary efforts of my predecessor and his team at U.S. Army Recruiting Command with great success in previous years, the Army had failed the recruiting mission just prior to my arrival in September 2005. So mission accomplishment during the following year in 2006 was an important milestone for the Army and the Nation.

Later that week after the stellar annual recruitment numbers had been released, I traveled to San Antonio. I was in my hotel room when my cell phone rang. I answered, "Hello, this is General Bostick, how can I help you?" The voice on the other end said, "Tom, this is Mike, Mike Mullen, how are you doing?" I answered, "Admiral Mullen, I am well, and you?" He went on to say, "Tom, I just wanted to call to thank you and the entire U.S. Army Recruiting team for your success this past year. We're all immensely proud of you and your team."

I thanked Admiral Mullen and hung up, somewhat in a state of shock that he reached out to me. Admiral Mullen was the Chairman of the Joint Chiefs of Staff. He was the most senior military officer in the Department of Defense and responsible for providing the best military advice to the President and the Secretary of Defense. Admiral Mullen was personally calling to congratulate my team and me. For me, it was a memorable example of relational leadership in action.

I have learned so much from leaders, soldiers, and civilians who I have known throughout my career, from the many leaders who I have served with and for, and I have also learned from the many who have worked for me.

EFFECTIVE RELATIONAL LEADERSHIP: MAKE IT PERSONAL

BE THERE IN PERSON

The story began when one of our soldiers, who was soon to receive an award, was driving from Georgia to his new duty assignment in Utah. He had spent several hours on the road with few breaks. This immobility over a long period of time caused him to develop a deep vein thrombosis, or a clot, which traveled to his heart and then to his brain, tragically causing a stroke. The outcome was devastating. He lost his ability to speak and was paralyzed on one side of his body. I knew about pulmonary embolisms, having experienced one myself. That's when I decided to present the award to him personally. When I arrived at the hospital, I was surprised to see the soldier in his uniform. As I approached the soldier, even though he struggled, he managed to stand up and salute. I returned his salute. After a few words, I presented his medal and gave him a big hug. There was not a dry eye in the room filled with doctors, nurses, and other patients.

But the story doesn't end there. Over a year later, I was invited to speak at an annual dinner gala. This same soldier who had lost his ability to speak and walk was at the podium giving the blessing for the dinner. Although speaking was difficult for him, everyone understood and appreciated his words. Later that night, with the support of his walker, he was on the dance floor with his wife enjoying the evening with so many others who, like me, were touched by his story.

REMEMBER PERSONAL CELEBRATIONS

Relational leaders have different techniques that allow them to focus on the individual while also achieving mission success. With respect to relational leadership, some of my practices focused on handwritten notes, visits, and many phone calls.

The U.S. Army Corps of Engineers has more than 34,000 people in the organization, including only seven hundred soldiers. Most people in the U.S. Army Corps of Engineers are civilians. The Army Corps has nine divisions, forty-three districts, and nine other units and labs. I had twenty-seven direct reports.

Given the number of people and organizations I wanted to remain connected to, we created a tracking system. We closely tracked which leaders I had met in my office or at a conference, those whom I had a purely "check-in" call with, and those organizations I had visited. We tracked the time since the last one of each of these engagements. Were the last engagements weeks, months, or over a year ago? Based on the length of time since the last engagement, we placed these leaders and organizations in categories of red, amber, and green. My executive team ensured that I focused on the red categories and did not miss special days like birthdays and anniversaries. My responsibilities had grown significantly since the days when my wife, Renee, would bake a double-layer chocolate cake for every soldier in the unit. But there was still a way to let people know that we were focused on the individual.

SEND PERSONAL NOTES

I also wrote star notes—thousands of them. A star note is a stationary card with a flag at the top displaying the number of stars representing the rank of the General Officer writing the card—in my case three stars. I have always believed that it is important to thank people, and I enjoyed sending personal handwritten notes. During my four years as the Chief of Engineers, I wrote over five thousand handwritten notes.

Months or years after writing a note, I would often meet someone who would remind me of a note that I had written to him or her. Sometimes they had framed the note as a reminder of my effort to reach out to him or her, and often they said they appreciated me thinking about the "little people." There are no "little people" on a team. It takes everyone on the team working together to achieve success.

In 2006, I signed an award for one of over 9000 recruiters. This particular soldier was a mid-career Non-Commissioned Officer. In 2020, this soldier reached the rank of Command Sergeant Major and was retiring from the U.S. Army. He contacted me to thank me for my leadership during our time together at U.S. Army Recruiting Command. He had never received one of my coins given to our best soldiers or one of my star notes. So, I sent a handwritten star note and a coin to him. Later, he sent me a picture of a retirement framed collage of his significant military accomplishments including my star note and coin.

Crisis management demands leadership that is not only tactical, strategic, and collaborative but also relational—leadership that listens, is empathetic, supportive, and responsive.

Building relationships is essential, and maintaining them is vital in both good times and bad, especially during a crisis. Remembering to reach out to check-in with your people personally using a variety of communication tools will resonate with them and help build strong winning teams especially during a crisis.

WHAT I LEARNED

1. **Relational leadership.** Crisis management demands not only leadership that is tactical, strategic, and collaborative but also leadership that is relational. That means leadership that listens and is empathetic, supportive, and responsive.
2. **Checking-in.** Reaching out to people through personal handwritten notes, birthday or anniversary phone calls, and individual and team recognition creates strong bonds in an organization in both good times and crises.
3. **Focus on people.** Taking the time to focus on people as well as on the mission helps build strong, winning teams.

You can't go back and make a new start, but you can start right now and make a brand new ending.
~JAMES SHERMAN

WINNING AFTER LOSING IS ALL ABOUT RESILIENCE

Life doesn't get easier or more forgiving,
we get stronger and more resilient.
~STEVE MARABOLI

THE RESILIENCE RESPONSE

Throughout this book, the central theme has been resilience and how to ensure that your leadership, your organization, and your teams have built not only a capacity for resilience but also understand how to draw on this critical resource on demand, and use it to maximum effectiveness. There are important questions to be asked and answers to be considered many weeks, months, even years before a person or an organization becomes more resilient. How does a person or an organization plan for the challenges ahead? How do they absorb an impact and bend but not break? How do they have a speedy recovery from a devastating impact? How do they adapt and become even stronger in preparation for a future challenge? How do they reimagine themselves? And what will the person and the organization look like because of this reimagining?

Resilience, the capacity not only to recover but to recover stronger through adaptation based on experience and lessons learned, is the cornerstone of this book. *Winning after Losing* is a book of stories about resilience—stories about how leaders and teams learned how to bounce back stronger after a loss, then learned to adapt and go on to win.

Crisis, challenge, and failure change people and their organizations. Resilient people, their leaders, and the organizations they are a part of never really "spring back into shape" but rather take on a new shape, a new form—stronger and better than before. This book addresses that important fact. Resilience, for those who understand how best to develop it, results in a new form of a person, of a leader, and of an organization that is better able to withstand stresses and challenges. Just as a bone after healing is stronger at the break, so is a resilient person, a resilient leader, and a resilient organization stronger once each has adapted to the crisis and emerged more resilient.

THE RESILIENT LEADER

The characteristics of resilience can be developed over time. And they can be learned through the combined leadership lessons that are described in the chapters of this book. What are the characteristics of a resilient leader? What skills does that leader need to guide an organization through a crisis or challenge and emerge healthier and stronger? While there are many characteristics identified in a variety of sources, *Winning after Losing: Building Resilient Teams* highlights the ten that follow:

1. **Personal well-being and focus on leadership health**: A healthy leader embodies the energy, stamina, and commitment needed and relies on a network of support to meet the demands of an ever-changing world.

2. **Self-efficacy**: Knowing that the leadership skills to succeed can be learned and applied; belief in one's ability to complete a task, assignment, or mission successfully.

3. **Self-confidence:** The ability to feel confident in oneself and one's decisions, and then to instill that confidence in others.

4. **Flexibility and Adaptability**: Rapid response in a critical situation; expediency in matching resources to needs.

5. **Risk-taking**: The ability to quickly assess risks and make difficult decisions.

6. **Realism**: Being able to make realistic yet flexible plans to overcome challenges.

7. **Communication**: Conveying a message that is aligned throughout the organization but also allowing for open, risk-free communication from others.

8. **Relational leadership**: Checking "in" rather then checking "on" members of a team, particularly during times of great challenge.

9. **Always learning**: Learning, adapting, and remaining open to innovative solutions strengthens resilient teams.

10. **Alignment on goals**: Aligning and communicating goals to achieve shared success throughout the organization.

PERSONAL RESILIENCE

I have had the privilege to experience the power of resilience on a personal level, not once, but repeatedly during my career. Whether it was finding Option 3 where a losing team found a way to win a major softball championship after suffering an initial embarrassing defeat; overcoming my own challenge of initially not receiving a Congressional nomination to West Point then bouncing back to secure a Presidential Nomination and going on to have a thirty-eight-year career in the Army; or bouncing back from a pulmonary embolism to deploy into combat and later compete in a Half-Ironman Triathlon; the key to success was always resilience. Individuals and teams must learn to become more resilient in the face of challenges, adversity, and crises.

Afterword

THE RESILIENCE LESSONS OF 9/11

I had the opportunity to support the Nation as the watch officer on duty in the National Military Command Center during 9/11. We worked closely with the President, Vice President, and the Secretary of Defense in responding to the terrible attacks on our Nation. As a nation, we adapted to become much stronger in our security and defense against a future attack. Our Nation adapted in many ways, such as the creation of the Department of Homeland Security, additional airport security requirements, locked airline pilot cabin doors, enhanced security to protect our borders, and increased entrance security requirements at all major venues including sporting events. There have been no similar large-scale terrorist events in the United States since 9/11. We became resilient. We bounced back stronger and better than we were before.

THE RESILIENCE LESSONS OF HURRICANE SANDY

Similarly, in the wake of Hurricane Sandy, the country became more resilient. For the first time, Congress appropriated funds for a collaborative study to review long-term methods for resilient adaptation to increased risks due to climate change. This study, the North Atlantic Coast Comprehensive Study (NACCS), provided a risk management framework that would support resilient coastal communities. There were many key findings, including the importance of encouraging resilience. I recall the days when Mayor Zimmer of Hoboken, New Jersey and I walked through his devastated city. The situation was dismal. Mayor Zimmer led one of the hardest-hit communities from Hurricane Sandy to bounce back in such a way that Hoboken was named as a role model for resilience by the United Nations. Hoboken did bend, but the city and its people did not break. Hoboken adapted and bounced back stronger and became more resilient.

THE RESILIENCE OF REACHING AN UNREACHABLE GOAL

As I recounted earlier, the Army had failed the recruiting mission in 2005 during a time when the Nation was fighting two wars in Afghanistan and Iraq. We adapted. We changed our slogan from "Army of One" to "Army Strong." We changed from "individual" to "team recruiting." We created a Medical Recruiting Brigade to focus on our medical professionals. We received local support from community leaders through a Grass Roots Community Advisory Board. We marketed the Army through NASCAR, the National Hot Rod Association, Professional Bull Riding, and the U.S. Army High School All-American Football Bowl Game. We bounced back from a tough recruiting loss in 2005 to winning in recruiting for more than a decade. We became more resilient.

RESILIENCE IN THE AFTERMATH OF THE VIETNAM WAR

During and after the Vietnam War, the military did not have the support of the American people. I told the story earlier that when I was a West Point cadet, my father, who was a Master Sergeant in the Army, warned me not to wear my uniform in public. The military had moved from a conscription force to an All-Volunteer Force. More than three decades after I entered West Point, I commanded the U.S. Army Recruiting Command. We made the decision to wear our combat uniforms, rather than our dress uniforms, in public. Regardless of where Americans stood on the wars in Iraq and Afghanistan, they supported their soldiers. Soldiers were thanked for their service everywhere they went. Gallup's rating of Americans' confidence in the military was 58 percent in 1974 when I entered West Point. Today, in the Gallup Poll that measures Americans' Confidence in U.S. institutions, the military stands at the top of the list of fifteen institutions that were assessed. The military is rated at 73 percent confidence. The military adapted and bounced back from the Vietnam War era to become the most respected institution in America. The military built its capacity for resilience over time.

THE RESILIENCE RESPONSE IN THE GLOBAL PANDEMIC OF COVID-19

The COVID-19 coronavirus has had a devastating impact on our Nation and the world. Our country had not seen a pandemic like the coronavirus since the Spanish Flu pandemic over one hundred years ago. The coronavirus has significantly impacted our economy, government, healthcare systems, education, and our way of life. As we absorb the impact of the coronavirus, we are bending but not breaking. We are learning to adapt in the face of trauma and tragedy, which has sadly become an integral part of our daily lives. Yet there is hope. We are a resilient nation, and despite this crisis and its costs, we will recover—stronger than ever before.

WHAT I LEARNED

1. **Resilience.** Resilience is the capacity to recover; it is the effort expended to bend but not break despite trauma, tragedy, adversity, or crisis. Ultimately, resilience is adapting based on the lessons learned, to bounce back stronger than before.
2. **Communications, teamwork, and alignment of goals.** The keys to success are teamwork and mutual support, clear and consistent communications, and alignment of goals at all levels that focus on the priorities critical to achieving success.
3. **America always bounces back.** From the founding of our Nation, we have a great history of bouncing back from difficult times. Whether it was the Spanish Flu, the Great Depression, two World Wars, terrorist attacks, or natural disasters, America has the talent, the resources, the leadership, and the will necessary to bounce back better no matter the challenge, adversity, or crisis.

> *Resilience: If you like winning, then learn*
> *as much as possible from losing.*
> ~LT. GEN. THOMAS P. BOSTICK

ACKNOWLEDGMENTS

First, I want to thank my wife, Renee, and son, Joshua, who have traveled the world with me during my Army career. They have sacrificed so much while contributing to my success. Renee has had a successful career in education, moved from school to school, state to state, from one continent to another. Joshua attended ten schools from kindergarten through the twelfth grade, yet still graduated as the Valedictorian of his high school and served as the golf team captain before going on to earn his undergraduate and graduate degrees from Stanford University. They both exemplified the definition of resilience throughout the many disruptions to their lives. Renee is also an expert editor. She was an editor of the U.S. Army War College proposal, a proposal that resulted in the accreditation of the U.S. Army War College, which then allowed students to graduate with a master's degree. Renee edited my Systems Engineering dissertation several times. And this book would not have been possible without her love, encouragement, and expert editing. Dr. Marie Kaye also supported me with my book. Marie provided astute insights and valuable recommendations that fine-tuned the book's content and organization. I also want to thank my executive assistant, Ms. Karen Huff, who has supported me in my military and civilian career for more than a decade. Karen was also extremely helpful in editing the book. Doran Hunter and the team at 1106 Design were enormously helpful in finalizing my book.

Next, I want to recognize my parents and siblings. I grew up with three brothers and one sister, and we traveled throughout Japan, Germany, and the United States with our father, Army Master Sergeant Sidney C. Bostick, and our mother, Fumiko M. Bostick. Our parents provided a loving and nurturing home and ensured that each of us had an opportunity to continue our education beyond high school—an opportunity that our parents never had. My siblings are Michael, an artist; Kathy, an attorney; Anthony, a veterinarian and Colonel in the U.S. Army; and my youngest brother Peter, a cancer surgeon.

During my career, I have remained in touch with many friends and mentors who have been a part of the stories in this book. I will be forever indebted to the United States Military Academy at West Point. The education and leadership experience at West Point was a once-in-a lifetime opportunity, and it has no equal. Playing on the 150-pound intercollegiate football team at West Point where each player could weigh no more than 158 pounds taught my teammates and me a lot about winning, losing, resilience, and friendship. Members of my West Point Class, "Proud and Great '78," are among my closest friends.

Lt. Gen. (Retired) Daniel W. Christman was our Battalion Commander during most of my time with Bravo Company, 54th Engineer Battalion at Wildflecken, Germany. General Christman is the type of leader who can work with presidents and heads of state, and hours later, one could find him giving a chest bump to his adoring troops or to cadets when he was Superintendent of the United States Military Academy at West Point. He and his exceptional wife, Susan, have been great friends throughout most of my career in the Army and in retirement. They were gracious and offered their home to Renee and me as the place to host our wedding reception at their home in Wildflecken, Germany, where we were stationed.

Major General Bruce K. Scott was our Executive Officer in the 54th Engineer Battalion during the maintenance success of Bravo Company. General Scott and his inimitable wife, Mary, have been lifelong friends. Generals Christman and Scott encouraged me to apply for a White House Fellowship, and I was fortunate to be selected. I have remained in regular contact with several friends from the 54th Engineer Battalion, including my first roommate in Wildflecken, Craig A. Crotteau, who is another lifelong friend; our Battalion Operations Officer, Gerry Hopkins; my company and Bravo Company Executive Officer, Steve Dunham, remain very close; and my West Point Classmate, Dick Thompson; Sergeant Major Scott R. Kuhar, who at the time was one of the junior soldiers in Bravo Company.

Brigadier General Peter Heimdahl and Colonel Edward G. Tezak provided a first-rate environment for professors in the Department of Mechanics, not only to teach the cadets but also to have fun in the process. The instructors were incredibly talented, and we have remained friends. Kathy Dennis, one of those instructors, is a dear friend and like family to us; we have served together in multiple assignments. After serving as the 165-pound athlete on an All-Army power lifting team, I coached the men's and women's powerlifting teams during my time at West Point as an instructor. We won the national championship the year after I departed from West Point. Paul Christopher was the coach, and I joined him and the team in Houston for the big win.

Serving as a White House Fellow was one of the most rewarding experiences of my life. As a White House Fellow, I served as a Special Assistant to Edward J. Derwinski, the first cabinet secretary of the Department of Veterans Affairs. Secretary Derwinski served as a Corporal in WWII. Each time I walked into his office, he stood, saluted, and would say, "Major Bostick, Corporal Derwinski reports." I had the opportunity to work with another former White House Fellow and Medal of Honor Recipient, Ronald E. Ray, and we have maintained a close friendship since our work together at the VA. I have remained in frequent contact with several of my 1989–1990 White House Fellows colleagues to include Leigh Warner, Gregory Hess, John McKay, Robert Marbut, John Orrison, and Wayne Tuan. Several of our class of White House

Fellows have re-connected each year during the annual White House Fellows Leadership Conference.

At U.S. Army Europe Headquarters in Germany, I had the opportunity to work for Brigadier General Bob Lee and then Colonel Joe N. Ballard, who would years later go on to serve as Lieutenant General Ballard, Commanding General of the U.S. Army Corps of Engineers. General Ballard was the first African-American to lead the U.S. Army Corps of Engineers. After my retirement from the military, General Ballard, a superb mentor and invaluable expert, assisted me with starting my own business. General Ballard and his lovely and peerless wife, Tess, have remained close friends.

Lieutenant General Randolph W. House was the Commanding General of 1st Infantry Division when I commanded the 1st Engineer Battalion at Fort Riley. General House taught leaders how to train and win on the battlefield. He mentored and counseled us in a clear, direct, and professional manner; he made us all better as warfighters and leaders. General House and his wife, Jeanie, led the 1st Infantry Division to great success. I had a top-notch group of officers and enlisted soldiers in the battalion who worked well as a team. Some of the soldiers became senior noncommissioned officers and some of the officers reached the rank of Colonel and General, going on to serve as successful civilian leaders.

In Germany, I worked for General Larry R. Ellis when he commanded the 1st Armored Division. General Ellis and his remarkable wife, Jean, led the division with great poise and confidence. General Ellis led the 1ˢᵗ Armored Division in Bosnia, where we worked very closely together. I have shared several stories in this book that highlight his vision, energy, and enthusiasm that led his teams to success. Several of his Brigade Commanders would go on to serve as General Officers to include Mark Kimmitt, Keith Walker, Volney Warner, and me. There were countless officers and enlisted soldiers who contributed to the division's success. Two officers who I would serve with multiple times were Brigadier General Dave Turner and Colonel Robert Sinkler. Both have been an integral part of my success at various levels.

I had many opportunities to serve in high-level staff assignments where I learned from outstanding leaders. Lieutenant General Arthur E. Williams, the 48ᵗʰ Chief of Engineers, selected me to become his Executive Officer. He would often say that his priorities were his family, his health, and his job, in that order—priorities, but also great advice. I also served on the Joint Staff in The Pentagon with Marine Lieutenant General Gregory S. Newbold, who was the J-3 responsible for worldwide operations, and with Major General Kip Ward, the Deputy J-3. General Ward managed the immediate operations during the tragic events of 9/11. He was always upbeat and positive even under the most trying circumstances. General Ward later served as the initial Commanding General of Africa Command.

I assisted with deploying the 1ˢᵗ Cavalry Division into Iraq. Major General Peter W. Chiarelli, an incredibly passionate warrior-leader, led the division in combat and later helped lead the U.S. Army as the Vice Chief of Staff of the Army. He worked vigorously on many important issues, including a primary focus on assisting our soldiers returning from combat with post traumatic stress (PTS). Beth Chiarelli did a tremendous job in leading the Family Readiness Group through some very challenging times. Several of our leaders would go on to help lead the Army as four-star Generals, to include Robert "Abe" Abrams at U.S. Forces Korea; Paul Funk II at Training and Doctrine Command; John "Mike" Murray at Army Futures Command; Edward M. Daly, at Army Materiel Command; and James McConville, Army Chief of Staff. There were also several other officers who deployed into Iraq with the 1st Cavalry Division, who were promoted to one-star, two-star, and three-star General Officers. Our Non-Commissioned Officers were simply outstanding, particularly our Division Command Sergeant Major, John Sparks, from whom I learned so much. We were fortunate to have a group of incredibly talented leaders at every level of the First Team.

Lieutenant General Robert "Van" Van Antwerp supported my assignment as the Commanding General of U.S. Army Recruiting Command. While this was a different type of assignment, it was one of the most rewarding of my military career. I worked with great civilian leaders who thoroughly understood recruiting, including Frank Shaffery and Rick Ayer. We had many superb officers and Non-Commissioned Officers who were resilient. We adapted after mission failure in 2005 to succeed year after year. The command was ripe for change and innovation. I recall asking General William S. Wallace if we could start a school for the Army that would award a General Education Diploma or GED to those high-quality young men and women that the Army could recruit. General Wallace told me to focus on recruiting, and he, along with Lieutenant General Ben Freakley, took on the responsibility of running the new Army GED school. While at recruiting, Dr. James Jones helped lead many changes in our effort to recruit medical professionals. He was our Physician Assistant, and he was an expert on all medical recruiting strategies. James has also been a friend and confidant, who has provided medical support to my family and me over many years. Colonel Tracy Cleaver and Colonel Jim Iacocca commanded the 3rd Recruiting Brigade during the transition to team recruiting. Colonel Gino Montagno was the first commander of the Medical Recruiting Brigade. These leaders were supported by an excellent team of officers, non-commissioned officers, and civilians.

As the head of personnel for the U.S. Army, I had the opportunity to work closely with the Assistant Secretary of the Army, Thomas R. Lamont. We worked well as a team, and Secretary Lamont was helpful in collaborating with key members of Congress to move along my nomination to become the 53rd Commanding General of the U.S. Army Corps of Engineers. I also worked closely with General George W. Casey, Jr. and General Chiarelli on many important areas for our military, including the repeal of the law commonly referred to as "Don't Ask, Don't Tell," increased opportunities for women, religious accommodations, and support for our wounded soldiers as they returned from combat. I relied heavily on support from Major General Jeffrey L. Arnold as we worked through these challenging issues. Jeff and his wonderful wife, Devon, have remained close and dear friends. My Executive Officers and assistants in the G1 were extremely competent and professional, to include Colonel Pat Gawkins, Colonel Steve Shappell, Colonel Larry Dillard, Colonel Gregory S. Johnson, and Ms. Karen Huff.

I had the great honor and privilege of working for General Eric K. Shinseki during his first two years as the Army Chief of Staff. General Shinseki led the Army through many innovative changes that successfully prepared the military for an uncertain future. General Shinseki and his incomparable wife, Patty, have known me longer than anyone else in my military career. I first met the Shinsekis at West Point when many other cadets and I found ourselves at their home on weekends. Throughout the years they have remained the same kind and gracious people since first meeting them so long ago. Other Army Chiefs of Staff that I worked with and learned so much from were Generals Peter J. Schoomaker, George W. Casey, Jr., Martin E. Dempsey, Raymond T. Odierno, and Mark A. Milley. General Casey and I worked closely together in Iraq and in The Pentagon. We discussed the law, "Don't Ask, Don't Tell," and he made the decision that the Army could support the repeal of the law.

As a General Officer, I have had the benefit of serving with a myriad of talented and dedicated Aides-de-Camp, Executive Officers, and Chiefs of Staff. A few that I wish to highlight here include: Robert Mihara, Gerald Austin, Catherine Rasch, Jon Clancy, Tony Prescott, Starr Corbin, Mike Rainey, Paul Hicks, Frank Myers, James Harris, Joel Holstrom, Rob Kellum, Gabe Marriott, Ricco Jones, James Welch, Lee Kemp, Jeff Johnson, Brian Bettis, Dave Ford, Cindy Atkins, Freddie Blakely, Renee Finnegan, Ronny Bagley, Pat Gawkins, Steve Shappell, Larry Dillard, Greg Johnson, Ken Reed, Jeff Hall, Whitney Hall, Andy Baker, Lisa "Reyn" Mann, Pete Andrysiak, Sheri Moore, Marc Hoffmeister, John "Chris" Becking, John Hudson, Dan Anninos, Dave Turner, Mark Toy, and Pete Helmlinger. I had several senior noncommissioned officers who I worked closely with in my senior assignments, to include: Jorge Gutierrez, John Sparks, Robert Hall, Jack Tilley, Harold Blount, Martin Wells, Stephan Frennier, Tom Gills, Karl Groninger, Antonio Jones, and Raymond Chandler. Master Sergeant Michael Young, our enlisted aide, was a master chef and close friend. Several Warrant Officers were key to my success in developing highly effective teams, including Dan Logan, Rich Alston, and Tony Prescott. All these Officers, Noncommissioned officers, and Chief Warrant Officers made me a better General Officer due to their leadership and support.

As I transitioned from the Army, several leaders were generous with their time and provided their valuable expertise. General Joe Ballard (49th Chief of Engineers, U.S. Army

Corps of Engineers) was key in guiding me in starting my own company; RJ Kirk (the CEO of Intrexon) gave me the wonderful opportunity to serve in the field of biotechnology. Dr. Tom Reed (Chief Science Officer of Intrexon) taught me so much about the scientific details of our business; and Dr. Manus Kraff became a wonderful friend and advisor. Joe Reeder and Kate Boyce-Reeder were tremendously helpful as we transitioned from the military. Sid Goodfriend assisted me in the Army and following my retirement. I received support from several colleagues while earning my PhD including Dr. Igor Linkov, Dr. James Lambert, Ms. Cate Fox-Lent, Dr. Jim Gigrich, Dr. John Domenech, Dr. Alberto Hernandez, Colonel Mary Lou Hall, Dr. Andrew Hall, Dr. Michael Deegan, Robert Sinkler, and my GWU advisors Dr. Thomas Holzer and Dr. Shahryar Sarkani.

Throughout my career, I have learned from these leaders, soldiers, and a host of many others. To those who touched the lives of the Bostick family and made us better because we knew you, we are ever grateful.

NOTES

1. Simon Sinek is a motivational speaker and the author of several books that highlight the importance of teamwork. While in the military, I had the opportunity to meet and listen to Simon Sinek several times. When Simon Sinek visited my team at the U.S. Army Corps of Engineers, he brought autographed books of *Leaders Eat Last*. This book focuses on Simon Sinek's experience with the Marines where he saw leaders eating last, which exemplified the fact that leaders sacrifice their own comfort for the good of those who are in their care. Simon Sinek has written and spoken frequently about why some teams succeed and others, with all the incentives to win, often fail.

2. The U.S. Army in Europe (USAREUR) was formed in 1942 and continues to operate from its headquarters in

Germany. During the 1980s, there were two U.S. Army Corps in Germany, the V Corps and the VII Corps. USAREUR and its allies had to be prepared for a potential assault into then West Germany. The timeline was short to respond to an attack by the former Soviet Union, and thus maintenance of vehicles was of such critical importance. The V Corps Maintenance and Assistance Inspection Teams (MAIT) conducted unannounced inspections to ensure that the vehicles and equipment were prepared to move on very short notice.

3. My Battalion Commander, for much of my time at the 54th Engineer Battalion, was then Lieutenant Colonel Daniel W. Christman, who later became a Lieutenant General serving as the Superintendent of the U.S. Military Academy at West Point, New York. He created an environment where leaders could freely communicate, learn, and win. Our Executive Officer was Bruce K. Scott, and I would work with him several times throughout my career. Both Generals Christman and Scott encouraged me to apply for a White House Fellowship. Steve Dunham was my Executive Officer in B Company where we achieved so much together with our team.

4. A-players are generally the most talented and focused athletes, as opposed to B-players, who are good players, but not the best.

5. B Company, 54th Engineer Battalion won the first Department of the Army Annual Unit Maintenance

Award in the intermediate category. *Engineer Magazine,* Volume 13, Number 1, 1983, Page 4.

6.   *Good to Great: Why Some Companies Make the Leap and Others Don't,* by Jim Collins. https://www.jimcollins.com/.

CHAPTER 2   THE MAILMAN DELIVERS . . . ONCE AGAIN

1.   The 1st Infantry Division, known as the Big Red One, is the oldest continuously serving division in the Regular Army. The Division Commander during my assignment with the division was Major General (later Lieutenant General) Randolph W. House, who was one of the greatest trainers of leaders and units I had ever served with in the Army.

2.   As part of the recruiting effort, the U.S. Army hosted the U.S. Army High School All-American Bowl Football Game. Many of these athletes went on to play in the National Football League, including Andrew Luck, Reggie Bush, Mark Sanchez, Adrian Peterson, Odell Beckham Jr., Tim Tebow, DeMarco Murray, and many more. Although the U.S. Army no longer sponsors this game, past Bowl games were held in January in San Antonio at the Alamo Bowl.

CHAPTER 3   WINNING IN SPORTS. WINNING ON THE BATTLEFIELD.

1.   The 1st Infantry Division Commander, Major General Randolph W. House, knew the value of competition and

the lessons gained through athletic competition. He led the 1st Engineer Battalion run before he presented me with the Commander's Cup Trophy in April 1996. *Fort Riley Post*, May 3, 1996, Page 10.

2. The 11th Armored Cavalry Regiment serves as the Opposing Force (OPFOR) at the National Training Center (NTC) and has the mission to train visiting units throughout the year.

3. A task force is a "temporary grouping of units, under one commander, formed for the purpose of carrying out a specific operation or mission." https://fas.org/irp/doddir/dod/jp1_02-april2010.pdf, page 467. My Brigade Commander created Task Force Bostick to serve as an additional maneuver task force. The forming of this task force was a unique approach at the National Training Center, but it was highly effective.

4. A hasty defense is "normally organized while in contact with the enemy or when contact is imminent and time available for the organization is limited." https://www.militaryfactory.com/dictionary/military-terms-defined.asp?term_id=2395.

5. A deliberate defense is "normally organized when out of contact with the enemy or when contact with the enemy is not imminent and time for organization is available." https://www.militaryfactory.com/dictionary/military-terms-defined.asp?term_id=1587.

6. "The Die Is Cast," Bob Knotts and Michael E. Young, *Sun-Sentinel*, May 29, 1994. https://www.sun-sentinel.com/news/fl-xpm-1994–05–29–9405310156-story.html.

7. The purpose of the Army Achievement Medal is to recognize the contributions of junior officers and enlisted personnel. https://www.federalregister.gov/documents/2006/04/05/06–2854/decorations-medals-ribbons-and-similar-devices. The fast actions of a junior Lieutenant to call for fire and destroy the Opposing Forces' Combined Arms Reserve earned him an impact Army Achievement Medal.

8. On December 16, 1944, the German Army launched a counteroffensive to cut through the Allied Forces. The battle that ensued is known historically as the Battle of the Bulge. The American soldier was tested and, through great adversity, prevailed. Never again was Hitler able to launch such an offensive against the West on such a large scale. https://www.army.mil/botb/.

CHAPTER 4   FROM MISSION IMPOSSIBLE TO MISSION ACCOMPLISHED

1. Major General Larry R. Ellis commanded the 1st Armored Division in Bad Kreuznach, Germany. He led the division in its deployment into Bosnia, and later became a four-star General and Commander of Forces Command, where he led the largest United States Army command and provider

of expeditionary, regionally engaged, campaign-capable land forces to Combatant Commanders. https://www.forscom.army.mil/.

2. The Davidson-Style SEA hut was named after 2nd Lieutenant Ross A. Davidson. This new design helped to maintain unit integrity, and it was the first change in design of the SEA hut since the Vietnam era. https://www.nato.int/KFOR/chronicle/1999/chronicle_199901/p12.htm.

CHAPTER 5 THE PORT OF RIJEKA, TUZLA AIRFIELD, AND KOSOVO

1. The Port of Rijeka is a seaport located in Rijeka, Croatia. Rijeka is the largest port in Croatia. During the 1990s, the port suffered a period of stagnation due to the Croatian War of Independence. The first vehicles from Fort Hood, Texas, arrived at the Port of Rijeka, Croatia on August 23, 1998. https://www.defense.gov/observe/photo-gallery/igphoto/2002017306/.

2. Tuzla is the third-largest city in Bosnia and Herzegovina, and it served as the location for the Headquarters of the 1st Armored Division.

3. Master Sergeant Patrick Daize was serving at Ramstein Air Force Base in Germany when I learned that he had the expertise to assist our team with building a strategic airfield at Tuzla, Bosnia. We were building a strategic airfield so that the troops from Fort Hood, Texas, could

fly directly into Tuzla. Patrick is a good example of what Jim Collins, in *Good to Great*, referred to when he wrote about getting the right people on the bus, and then getting the people in the right seat on the bus.

CHAPTER 6   ON THE POWER OF CALM AND QUIET LEADERSHIP

1.   "How Humble Leadership Really Works." Both General Eric K. Shinseki and General Fred M. Franks are humble servant leaders. They live the words "Mission First, People Always," and possess incredible emotional intelligence in times of great challenge during peace and war. Dan Cable, "How Humble Leadership Really Works, " *Harvard Business Review*, April 30th, 2018. Retrieved from https://hbr.org/2018/04/how-humble-leadership-really-works.

2.   George Will wrote a wonderful article entitled, "Young Sailor's Story Provides Good News That Heroes Still Exist," which appeared in the *Baltimore Sun* on October 21, 1999. Very much worth reading. https://www.baltimoresun.com/news/bs-xpm-1999–10–21–9910210353-story.html.

3.   The Hall of Valor Project. https://valor.militarytimes.com/hero/34356.

CHAPTER 7   FIRST TEAM

1.   History of the 1st Cavalry Division. https://1cda.org/history/. General Peter W. Chiarelli was the Division

Commander when the 1st Cavalry Division deployed into Iraq in 2005. I assisted with the division's deployment into Iraq from Fort Hood and Kuwait. I returned to Iraq and served closely with General Chiarelli in rebuilding the infrastructure of Iraq when I was the Commander of the Gulf Region Division. General Chiarelli went on to become the Army Vice Chief of Staff, the second highest military position in the Army. I worked closely with him in The Pentagon when I was responsible for Army Personnel Policy and Resources.

2. History of the 4th Infantry Division. https://www.4thin-fantry.org/content/division-history. General Raymond T. Odierno was the Division Commander of 4th Infantry Division. He later became the Army Chief of Staff. I worked closely with him in The Pentagon when I was responsible for Army Personnel Policy and Resources, and later as the Commander of the U.S. Army Corps of Engineers.

3. The 4th Infantry Division attack into Iraq changes from the north through Turkey to the south through Kuwait. https://www.foxnews.com/story/mighty-4th-infantry-finally-enters-southern-iraq.

4. The National Training Center (NTC) at Fort Irwin, CA provided U.S. Army units the opportunity to conduct demanding realistic training during a twenty-one-day cycle prior to deploying overseas.

5. The After-Action-Review (AAR) is a powerful military and business tool. M. Darling, C. Parry, and J. Moore,

"Learning in the Thick of It," *Harvard Business Review*, August 1ˢᵗ, 2014. Retrieved from https://hbr.org/2005/07/learning-in-the-thick-of-it.

CHAPTER 8   THE POWER OF PERSEVERANCE

1.   Major General David H. Petraeus commanded the 101ˢᵗ Airborne Division in Iraq and later served as the Commanding General of Central Command. General Petraeus has a reputation for his superb physical fitness in addition to his great skills as a soldier, scholar, and leader.

2.   Sergeant Tony Prescott supported my team when I was the Assistant Division Commander for Support in 1ˢᵗ Cavalry Division. He worked hard to ensure I took some time off away from work. He once ensured that I went deer hunting, and I was successful. Tony was in the ambulance when I was taken to the hospital for an angiogram. Our families have become lifelong friends. Sergeant Tony Prescott later became Chief Warrant Officer Tony Prescott.

3.   Carl R. Darnall Army Medical Center is the hospital at Fort Hood. I'm alive today because of the great work by our military and civilian medical professionals who support our troops and families with the best medical care possible. Dr. James Jones, the Physician Assistant in U.S. Army Recruiting Command, has provided my family and me with the best medical care possible.

My ability to recover from my pulmonary embolism and to train and compete in athletic competitions is a great testament to his support. James went on to be the Physician Assistant for President Obama and President Trump.

4. Deep vein thrombosis and a pulmonary embolism can be deadly. A reporter and co-anchor of *Weekend Today*, David J. Bloom died suddenly after a deep vein thrombosis (DVT) became a pulmonary embolism. David Bloom was traveling with the 3rd Infantry Division in cramped conditions in a military vehicle during the initial attack toward Baghdad, Iraq, on April 6, 2003 when he died. https://www.today.com/allday/10-years-later-david-bloom-remembered-1B9239813.

5. The Warfighter Exercise is training focused on developing core warfighting competencies in conjunction with the unit's training objectives. https://usacac.army.mil/organizations/cact/mctp.

6. Specialist Ray Joseph Hutchinson was a great soldier, a wonderful son, and a man of faith. His parents, Michael and Deborah Hutchinson, started the Ray Joseph Hutchinson Foundation (RJH Foundation) to help recognize the service and sacrifice of their son. Along with other soldiers, I have remained in touch with the Hutchinson Family, on difficult but special days where we think about Ray Joseph. https://www.facebook.com/RJHFoundation/.

CHAPTER 9   TURNING DIRT

1. General George W. Casey Jr., 36th Chief of Staff of the Army. I served with General Casey in Iraq and when he was the Army Chief of Staff. After a smooth transition with my good friends Rear Admiral David J. Nash and Major General Ronald L. Johnson, I worked with General Casey, Ambassador William Taylor, and Mr. Charlie Hess to significantly expand the construction program.

2. William D. Wunderle, *Through the Lens of Cultural Awareness: A Primer for U.S. Armed Forces Deploying to Arab and Middle Eastern Countries* (Combat Studies Institute Press, 2006), V.

3. Special Immigrants Visas (SIVs) for Iraqi and Afghan Translators/Interpreters. https://travel.state.gov/content/travel/en/us-visas/immigrate/siv-iraqi-afghan-transla-tors-interpreters.html.

4. Tsedal Neeley, "Global Teams That Work." *Harvard Business Review*, October 2015.

CHAPTER 10   FIRST, BREAK ALL THE RULES

1. U.S. Army Recruiting Command (USAREC) is the Army organization with the mission to recruit for the Army. I served at USAREC for forty-four months of an originally planned eighteen-month assignment. My time

at USAREC was one of the most rewarding and most challenging of my career. https://recruiting.army.mil/.

2. The Medical Recruiting Brigade turned out to be an excellent idea. The Medical Recruiting Brigade not only accomplished its mission but gave an excellent command opportunity for future Army senior leaders. Lieutenant General Scott Dingle, the Army Surgeon General, commanded the Medical Recruiting Brigade. https://recruiting.army.mil/mrb/about_mrb/.

3. *First, Break All the Rules: What the World's Greatest Managers Do Differently,* by Marcus Buckingham and Curt Coffman.

4. Grassroots Community Advisory Boards contributed to the success of Army Recruiting. https://www.army.mil/article/54203/getting_america_involved_in_recruiting_its_army.

5. Civilian aides to the Secretary of the Army (CASAs) are business and community leaders who are appointed by the Secretary of the Army to help support the Army across the country. They are invaluable in supporting U.S. Army Recruiting Command. https://www.dcmilitary.com/pentagram/community/who-are-the-casas-what-do-they-do/article_2d37fa58-bd56–5e98-b87f-46a378d9cd47.html.

CHAPTER 11 OUR SOLDIERS

1. *Michigan Dad Loses 230 Pounds to Enlist in Army,* https://archive.defense.gov/news/newsarticle.aspx?id=18thirty-eight8.

2. *Why Are Southerners So Fat?* http://content.time.com/ time/health/article/0,8599,1909406,00.html.

3. *Obesity: Overview of an Epidemic.*

4. https://www.ncbi.nlm.nih.gov/pmc/articles/ PMC3228640/.

5. Leon Butler, New Orleans Recruiting Battalion, "Civilian Iraq Veteran Goes Army," *Recruiter Journal*, October 2006, page 10. Retrieved from https://archive. org/stream/recruiterjournal5810ftsh/recruiterjour-nal5810ftsh_djvu.txt.

6. "Angelo Vaccaro, A Real American Hero," https:// townhall.com/columnists/jerrynewberry/2006/10/16/ angelo-j-vaccaro,-a-real-american-hero-n1100587.

7. Military Hall of Honor, "Angelo Vaccaro," https://mili-taryhallofhonor.com/honoree-record.php?id=8565.

8. "Walter Reed Names Building After Fallen Combat Medic," https://www.army.mil/article/4297/wal-ter_reed_names_building_after_fallen_combat _medic.

CHAPTER 12   PANCAKES FOR DINNER STRATEGY

1. "'Army Strong' Replaces 'Army of One,'" http://www. nbcnews.com/id/15197720/ns/us_news-military/t/army-strong-replaces-army-one/#.Xnozd4hKiUk.

CHAPTER 13   THE POWER OF TEAM DIVERSITY

1. The 442nd Regimental Combat Team, with the motto
    "Go for Broke," was an Army unit composed of Japanese-
    Americans from Hawaii and the mainland of the United
    States. http://www.goforbroke.org/learn/history/mili-
    tary_units/442nd.php.

2. Terry Shima served with the 442nd Regimental Combat
    Team during World War II. https://memory.loc.gov/
    diglib/vhp/story/loc.natlib.afc2001001.56511/.

3. The Congressional Gold Medal has been bestowed by
    Congress to individuals since the American Revolution for
    distinguished achievements for the Nation. https://history.
    house.gov/Institution/Gold-Medal/Gold-Medal-Recipients/.

4. All combat roles were opened to women beginning in
    January 2016. https://www.defense.gov/Explore/News/
    Article/Article/632536/carter-opens-all-military-occu-
    pations-positions-to-women/.

CHAPTER 14   AT THE INTERSECTION OF LEADERSHIP AND PERFORMANCE

1. "H.R.2965—Don't Ask, Don't Tell Repeal Act of
    2010," https://www.congress.gov/bill/111th-congress/
    house-bill/2965.

2. Vice Chief of Staff of the Army, General Peter Chiarelli,
    led the Army Staff and was the second most senior officer

in the Army behind the Chief of Staff of the Army, who at the time was General George Casey.

3.   Joe McDade was my deputy when I led Army Personnel. He served an important role with the responsibility for planning the training required if the law, "Don't Ask, Don't Tell," was repealed, as well as the implementation of that training.

4.   The Honorable Jeh Johnson served as the Department of Defense General Counsel in addition to serving as the Co-Chair of the Comprehensive Review Group that analyzed the impact of the potential repeal of the law, "Don't Ask, Don't Tell." Jeh Johnson would later serve as the Secretary of Homeland Security.

5.   General Carter Ham served as Commander of United States Africa Command in addition to serving as the Co-Chair of the Comprehensive Review Group that analyzed the impact of the potential repeal of the law, "Don't Ask, Don't Tell."

6.   Among several responsibilities, Lieutenant General Mark Hertling oversaw the training for new soldiers, or basic training. He provided me with the letter stating that the Army could train Sikh soldiers without a disruption to good order and discipline.

7.   Specialist Simran Lamba, "Sikh Soldier Answers Lifelong Calling to Serve," https://www.army.mil/article/58866/sikh_soldier_answers_lifelong_calling_to_serve.

8.   Major General Jeffrey L. Arnold assisted me in thinking through some of the most challenging personnel matters

in the Army. He remains one of the most thoughtful leaders that I know. Colonel Pat Gawkins served as my Executive Officer during most of my time as the Director of Personnel, and I relied on him heavily, along with Major Larry Dillard, for their insights.

CHAPTER 15   THE ART AND SCIENCE OF IDENTIFYING FUTURE LEADERS

1.   Presidential Nomination to Service Academies. *Congressional Nominations to U.S. Service Academies: An Overview and Resources for Outreach and Management*, page 6. https://fas.org/sgp/crs/misc/RL33213.pdf.

2.   Brigadier General George D. Wahl graduated from West Point in 1917 and served in World War II with the 79th Division Artillery. He died on March 24, 1981, and was buried at Arlington National Cemetery. He was one of my earliest and most influential mentors who helped to launch me into a career of public service.

3.   Margaret Burcham was the first female Brigadier General in the U.S. Army Corps of Engineers (USACE), and the only other minority serving as a General Officer when I became Chief of Engineers. She inspired many other females, as well as male officers, who now serve at key leadership positions in the U.S. Army Corps of Engineers.

4.   Major Lisa Reyn Mann became my Aide-de-Camp for my final eighteen-months in the Army. She was a terrific aide.

CHAPTER 16 CRISIS MANAGEMENT AND LEADERSHIP HURRICANE SANDY: A CASE STUDY IN CHALLENGES AND SOLUTIONS

1. "Hurricane Sandy Fast Facts," https://www.cnn.com/2013/07/13/world/americas/hurricane-sandy-fast-facts/index.html.

2. National Response Framework. January 2008. https://www.fema.gov/pdf/emergency/nrf/about_nrf.pdf.

3. Emergency Support Functions. The U.S. Army Corps of Engineers is responsible for Emergency Support Function #3, Public Works and Engineering, https://www.phe.gov/Preparedness/support/esf8/Pages/default.aspx.

4. Congress Establishes the U.S. Army Corps of Engineers. https://www.history.com/this-day-in-history/congress-establishes-the-u-s-army-corps-of-engineers.

5. Dewatering the Brooklyn Battery Tunnel. https://www.nad.usace.army.mil/Media/News-Releases/Article/48thirty-eight27/dewatering-the-brooklyn-battery-tunnel/.

CHAPTER 17 LEADERSHIP AND PEOPLE—A WINNING COMBINATION

1. Schofield's Definition of Discipline. https://www.themilitaryleader.com/quotes/schofield-on-discipline/.

2. Aide-de-Camp. I had a total of twenty-five Aides-de-Camp. I used the position not only to assist me, but

to assist the Army in developing talent. https://www.britannica.com/topic/Aide-de-Camp.

3. Talent management in business is possible. Grey Frandsen, CEO of Oxitec near Oxford, England, was among our best leaders in the company. I regularly provided him with opportunities to participate and gain insights into the planning and execution at the corporate level. These opportunities gave him a much better understanding of the big picture, and he proved to be a superb strategic leader.

CHAPTER 18 PUBLIC-PRIVATE PARTNERSHIPS

1. U.S. Army Corps of Engineers has a key responsibility for flood risk management for the Nation. https://www.usace.army.mil/About/History/Brief-History-of-the-Corps/Improving-Transportation/.

2. The Residential Communities Initiative resulted in homes that were built, maintained, and managed by the private sector. All services of the military participated in this program that provided quality housing on the military bases for the service members and their families. https://www.rci.army.mil/.

3. Rand Study, *Military Installation Public-to-Public Partnership*. The "Monterey Model" demonstrates how the City of Monterey used its public capabilities to partner with another public entity, the Naval Post Graduate School, to develop a win-win scenario. https://www.rand.

org/content/dam/rand/pubs/research_reports/RR1400/
RR1419/RAND_RR1419.pdf.

4. Fargo-Moorhead Public-Private Partnership. "Army
    Corps of Engineers' First P3, a $2.8B Flood Control
    Project Restarts." https://www.constructiondive.com/
    news/army-corps-of-engineers-first-p3-a-28b-flood-con-
    trol-project-restarts/560588/.

CHAPTER 19   TALENT MANAGEMENT

1. The White House Fellows Program was founded by
    President Lyndon B. Johnson in 1964. The White House
    Fellowships offer exceptionally talented men and women
    an opportunity to serve at the highest levels of the federal
    government. The White House Fellows program is one
    of the nation's most prestigious programs for leader-
    ship and public service. https://www.whitehouse.gov/
    get-involved/fellows/.

2. U.S. Army Recruiting Command recruits the men and
    women who will serve the Nation in the Active Army and
    the Army Reserve. https://recruiting.army.mil/.

CHAPTER 20   YOUR PEOPLE ARE YOUR BEST BRAND AMBASSADORS

1. The cadets of West Point supported the making of the
    movie MacArthur. https://moviesanywhere.com/movie/
    macarthur.

2. General MacArthur's "Duty, Honor, Country" speech to West Point cadets, May 12, 1962. http://www.macarthurmilwaukeeforum.com/resources/macarthurs-speech-to-west-point-cadets-may-1962/.

3. Anheuser-Busch "Applause" commercial shows members of the U.S. Armed Forces walking together through an airport terminal. One person starts applauding the troops, and this action suddenly turns into the entire terminal standing and clapping as the men and women in uniform walk by. This 2005 Super Bowl commercial was hugely successful. https://6ammarketing.com/blog/anheuser-busch---applause.

CHAPTER 21   FROM BOSS TO MENTOR TO FRIEND: THE POWER OF RELATIONAL LEADERSHIP

1. J. Coleman, "The Power of Relational Leadership," *Forbes*, August 18th, 2018. Retrieved from https://www.forbes.com/sites/johncoleman/2018/04/16/the-power-of-relational-leadership/#20f8d0c1369d.

2. T. Butler and J. Waldroop, "Understanding 'People,'" Harvard Business Review, August 1st, 2014. Retrieved from https://hbr.org/2004/06/understanding-people-people.

3. Baltimore Mediation, "The 5 Components of the Relational Leadership Style Of Leading," November 30th, 2018. Retrieved from https://www.baltimoremediation.com/5-components-relational-leadership-style-leading/.

WHAT I LEARNED, CHAPTER BY CHAPTER

CHAPTER 1 DISCOVERING OPTION 3

This chapter highlighted a losing team that was turned into a winning team through sports, travel, and the power to make an "out of the box" decision. It is the chapter that defines the concept behind this book.

WHAT I LEARNED

1. **Allow your team to offer alternative options**. Having different options is a recognized leadership skill; however, what I learned was to let your team come up with some options of their own and implement those options as alternatives. You may find that sometimes the best path to a winning outcome may not be the most direct. This

is a leadership strategy that I have since used frequently to good effect.

2. **No need for all A players to build a championship team.** All too often leaders feel they can't build outstanding performance-driven teams unless they have a team made up of only star players, A players. I found this not to be the case. I had few A players, and even so, managed to build a championship team in both sports and maintenance. A big part of winning for any team is teamwork and most of all trust.

3. **Trust.** Even if a team is made up of a seemingly random mix of people with very little in common, they can bond strongly and develop interpersonal trust.

4. **Let individual talents shine.** Even though a team may be focused on one single goal, each member of that team has different skills to contribute. A team does best when each member is placed in a position where their individual skills and talents shine. Jim Collins highlighted a similar concept in his book, *Good to Great: Why Some Companies Make the Leap and Other's Don't,* in which he stressed the importance of getting the right people on the bus, then getting these people in the right seat on the bus, before trying to figure out where the bus is going. Although *Good to Great* was published well after we developed a championship softball and maintenance team, we used the same concept as the bus analogy.

5. **Downtime.** Many of us work in high-performance settings as do our teams. But it is important to keep in mind that a team needs to have some downtime or a non-work-related activity (in my case it was softball and traveling Europe) to help bring the team together.

CHAPTER 2   THE MAILMAN DELIVERS . . . ONCE AGAIN

This chapter focused on a single person, in this case NBA Basketball Star, 'The Mailman,' Karl Malone that demonstrated how a little help in winning can come from an unexpected source.

WHAT I LEARNED

1. **Be agile and open.** Being agile and open means being prepared through creativity and a positive mindset to accept help from unexpected sources.

2. **Great training has universal applications.** A well-trained team can win battles on any terrain and against any enemy. Similarly, well-trained teams of civilians can adapt and win in a completely different area with the proper training, resources, teamwork, and leadership.

3. **Don't dismiss the unconventional.** Unconventional tools, like the footballs and frisbees that our team used, may be the very glue that binds a team together.

4. **Appreciate the power of sports**. Sports play a large part in forging and strengthening many different types of teams in more cohesive and effective units.

CHAPTER 3   WINNING IN SPORTS. WINNING ON THE BATTLEFIELD

In this chapter, sports and healthy competition are discussed as powerful tools for bonding teams while still maintaining alignment with the larger organization.

WHAT I LEARNED

1. **Sports can be a powerful team building tool**. Sports can be a very powerful team-forging tool and if used correctly can not only help teams achieve their goals but also can build overall confidence and boost motivation.

2. **Competition**. Competition within an organization can be a powerful motivator and driver for successful goal achievement.

3. **Certification is important in order to train and build confidence in individual teammates.** These certified leaders will believe in themselves because they know their craft and can execute under pressure.

4. **History**. History, if used judiciously, can also be a significant source of inspiration. Remembering and reliving the achievements and successes of the past can drive desire for success in the present and in the future.

CHAPTER 4 FROM MISSION IMPOSSIBLE TO MISSION ACCOMPLISHED

In this chapter, leaders exemplify the true meaning of vision and how to inspire teams to success against all odds.

WHAT I LEARNED

1. **Leaders with vision.** The first lesson is one that was well expressed by American philosopher and scholar Charles F. Kettering, "A problem well stated, is a problem half-solved." Leaders with vision who can state that vision clearly, like Major General Larry Ellis, who wanted housing that would go beyond simple shelter but would serve as a life-saving team-building tool, stated his vision clearly.

2. **Look for fresh ideas from the young and inexperienced.** The second lesson is that the young and inexperienced have a lot to offer. They often provide fresh ideas, turn existing models inside out and upside down, and have the energy to execute those ideas and turn them into workable solutions. Leaders must create an environment that allows all members of the team to feel comfortable in speaking up—especially those least experienced. The secret sauce that the best leaders add to their mix is trust. In a climate of trust, even the most junior team member feels that their contribution will be valued and considered with respect.

3.   **Respect individual passion:** The third lesson learned is that while a team is a powerful unit, it is made up of individuals who must make their own decisions regarding their path in life. And if an individual has a passion and does the necessary homework then they will succeed on any team—just as the young, untried, but talented 2nd Lieutenant Ross Davidson made an impossible mission possible.

CHAPTER 5   THE PORT OF RIJEKA AND TUZLA AIRPORT

This chapter celebrates the art of leadership vision and setting high goals that serve as inspiration and achievement success.

WHAT I LEARNED

1.   **Raise the bar.** I learned the power of setting high expectations. Leaders can and should challenge the status quo and the boundaries set by the past when forging a new future.

2.   **Ask the right questions.** Asking the right questions can make all the difference between success and failure. Getting to the very heart of the challenge, deciding what's needed, and then providing the appropriate resources can result in success.

3.   **Visionary leadership.** Leaders who have vision are critical to success, especially in disruptive times. These leaders

not only envision bold solutions but envision their teams successfully executing their vision. Visionary leadership leads not only to successful missions but to motivated and confident teams.

4. **Experts.** Expertise is invaluable when tackling challenging missions on a very short timeline. Seek out and find the experts and add them to your team.

CHAPTER 6   ON THE POWER OF CALM AND QUIET LEADERSHIP

This chapter reinforces the importance of focusing on the people under your leadership before everything else—even preparation for a Congressional hearing.

WHAT I LEARNED

1. **Leadership**. Leadership exists at all levels of experience and can be found in the youngest among us.

2. **Role model**. Being a role model is not linked to seniority and rank. Senior and experienced leaders can clearly serve as great role models who inspire the young; however, young people such as Ensign Johnson are equally inspiring role models.

3. **Resilience.** Ensign Johnson displayed the power of resilience and how someone can experience a tragic life changing event, recover, and bounce back as a strong and successful person.

4. **Modesty**. When one lives a life of not taking credit, credit never leaves you.

In this chapter, the importance of reinforcing the alignment throughout all levels of an organization even while encouraging competition between corporate teams.

WHAT I LEARNED

1. **The value of organizational alignment at all levels.** Rivalry between teams can be healthy if the core bonds of unity are not broken and if alignment throughout the organization is placed first above all else.

2. **The After-Action-Review (AAR).** The After-Action Review is a powerful tool for teams. It allows all members of the team to communicate openly, regardless of rank or standing. This results in a better understanding of what went wrong in an operation, why, and how to fix key factors in the future so that the same mistakes are not repeated in the future. The AAR is an excellent tool for the military and business alike.

3. **Making seemingly counterintuitive decisions.** Leaders must sometimes make the tough call that may seem counterintuitive but is actually 'right on.' Leaders must often go with their instincts honed over years of testing.

4.   It is important to remember the families of team members. Families of team members sacrifice as much as those who serve in the military or in business.

CHAPTER 8   THE POWER OF PERSEVERANCE

This chapter discusses the vulnerability of leaders and how important it is for them to stay strong and healthy, not only for themselves, but the organizations they lead and the teams that depend on them.

WHAT I LEARNED

1.   **Personal focus.** Leaders must focus on their own personal resilience and pay close attention to their own mental and physical fitness by preparing their mind and their body for setbacks.

2.   **Helping others when you are down.** A leader can have a bad day, but not in front of the troops. Even if a leader is having a difficult day, the troops will continue to require great leadership. Supporting others will give renewed strength to the leader. Booker T. Washington said, "Those who are happiest are those who do the most for others."

3.   **Selfless service.** As Ray Joseph demonstrated, many soldiers in combat are concerned most about their fellow soldiers. That is what strong teammates do. In business,

leaders should consider what sacrifices teammates make, and why they make them. Is the sacrifice for their fellow workers, the company, or something else?

4. **Accept support.** When knocked down, learn to accept support from family, friends, and colleagues. In this way, leaders can find themselves winning after losing due to a health setback.

5. **Bounce back strong.** After experiencing a setback, it is important to bounce back stronger than before with even greater mental and physical readiness as well as the skills to employ protective measures. Personal resilience is developed over time by preparing for setbacks through mental and physical training, absorbing the impact by bending but not breaking, recovering properly, and then adapting to become stronger than before.

CHAPTER 9   TURNING DIRT

For leaders whose organizations are global, this chapter offers several illustrations of how to lead a global organization along with its global teams.

WHAT I LEARNED

1. **Cultural awareness.** It is crucial to listen to people who may know best what is happening on the ground, who are familiar with the local culture and its customs. It is

then crucial to adapt your own thinking and actions to succeed in that new culture.

2. **Two kinds of team players.** In today's global economic and organizational context, teams are frequently made up of both internal organizational members as well as external players. It is important to have an inclusive team atmosphere in which external team members feel important and their expertise is valued.

3. **Go with your instincts.** Sometimes the most efficient and cost-effective solution will not succeed. Know when to change. If one effort fails, modify, and try again. Often a leader must go with their instincts. Know when to pivot to an approach that may not be the most direct but may be more successful.

CHAPTER 10   FIRST, BREAK ALL THE RULES

This chapter highlights the necessity for leaders to sometimes start from the ground up, take risks, and do what is ultimately best for the organization no matter in what department.

WHAT I LEARNED

1. **The best time to change is often just after being defeated.** The key to winning after losing is recognizing that this is often the right time to move in a new direction regardless

of how challenging it may be. It is important not to stay with a bad decision but to put it behind you and create a new path.

2. **Be creative in the composition of your team.** Think outside the organization when you are building your teams. The Grassroots Community Advisory Board reinforced for me that the Army team is broader, including not only uniformed soldiers but also patriotic civilians.

3. **Speak with one voice.** It is important when leading a team or an entire organization to ensure that your message is consistent and internalized. Effective leaders and everyone on their team must speak with one voice.

4. **Listen to your people in the trenches.** Those junior people in the trenches, who must execute policies, could very well know best how to succeed in their unique environments assignments. Keep an open mind.

5. **Set aside time to reflect and assess changes that have been made.** Recognize when the change envisioned, planned, and executed is not working as originally desired and don't be afraid to admit defeat and change again. Change is constant.

6. **Be intentional.** Go see for yourself. Ask questions. Listen. Make mental notes. Take written notes. Celebrate success.

CHAPTER 11   OUR SOLDIERS

Heroes in organizations are often those who set high personal goals and reach them. Heroes overcome great personal odds—they are the faces of those who know the joy of "winning after losing."

WHAT I LEARNED

1. **Never give up.** The stories of courage and tenacity in this chapter show how important it is for leaders to instill into their teams the idea of never giving up on your dreams. A great deal can be achieved by those who work in an environment of support and encouragement.

2. **Go as far as you must go to make positive change happen.** Sometimes you surmount many obstacles but always keep moving forward. Go as far as you need to go to find your dream, to make a difference, and to effect change whether in the military, in business, or government.

3. **Heroes are often ordinary people doing extraordinary things.** In each of these stories, it was the personal drive and belief in themselves that helped each of these heroes achieve his or her dreams. Will power, perseverance, and support from others made all the difference.

CHAPTER 12 PANCAKES FOR DINNER STRATEGY

Organizations do not just owe their success to their teams or even to their leadership. They also owe their success to the silent army of supporters—family, friends, and the quiet ones—who all contribute.

WHAT I LEARNED

1. **Recognize the wider circle of contributors.** Family and friends play a significant role in the successful retention of the members of an organization. Support from spouses, family members, and friends can make a difference for most organizations. Family and friends are often part of what makes an organization strong and can influence whether a member remains with or leaves the organization. Consider how your organization recognizes these important members of the team who are not on the payroll.

2. **Pay attention to the quiet team members.** Leaders need to pay attention to the humble and quiet members of their teams. As they say, "The squeaky wheel gets the grease." Leaders must ensure they are mindful of those in the organization who are modest, unassuming, and uncomplaining. These people may not advocate for themselves but expect that leaders will.

CHAPTER 13 THE POWER OF TEAM DIVERSITY

It has been shown again and again that diversity makes an organization stronger and counts a great deal toward its success in today's world.

WHAT I LEARNED

1. **Recognizing our heroes.** It is never too late to recognize heroes. It is never too late to address the challenges of the past.

2. **Diversity.** There is great strength in diversity of gender, ethnicity, religion, and ideas. Diversity makes the military and civilian organizations stronger.

3. **What counts in assessing talent.** Performance is the key metric.

CHAPTER 14   AT THE INTERSECTION OF LEADERSHIP AND PERFORMANCE

A further look at the challenges as well as opportunities brought out by embracing diversity in organizations.

WHAT I LEARNED

1. **Supportive leadership.** When you know a leader has your back, you are stronger, more confident, and ultimately more successful.

2. **The leader also needs support.** It can be very lonely at the top of any organization. Especially during challenging

times, even a leader needs support. Reaching out to a leader by way of a simple phone call will go a long way.

3. **The importance of listening and gathering data.** Any significant change for an organization will have people on both sides of an issue. Listening, gathering information, and reviewing data and evidence are invaluable in the decision-making process.

4. **Diversity is strength.** The strength of the military or any organization is in its diversity.

5. **It is all about performance.** Rather than focusing on what distinguishes different groups, focusing on performance will be a unifying factor in building great teams.

6. **Take on challenges when you feel least prepared.** Sometimes taking on challenging matters should occur when the demands on an organization are the greatest.

CHAPTER 15 THE ART AND SCIENCE OF IDENTIFYING FUTURE LEADERS

This chapter is about so much more than having or being a mentor. It is about the importance of mentorships in the success and continuation of virtually every organization and its leadership.

WHAT I LEARNED

1. **Start with yourself.** Leverage your passion and leadership style so that you best understand how to assist others.

2. **Find a role model.** Role models can offer a glimpse into the promise the future holds.

3. **Find your passion**. It has been said that if you love your job, it's not really work. However, finding your passion may not be simple. It may develop over time. It requires discovery—knowing yourself, your talents, then matching them to a life's work.

4. **Be persistent.** There is generally always a way to reach your objective. One just needs to find the right path.

5. **Be ready for an unexpected mentor**. A mentor could be someone you don't know, but who has an interest in your success. Be open to offers of guidance, advice, and support.

6. **Become a role model.** One does not need to be older, wiser, and more experienced to serve as a role model.

7. **Find the specific talent you need.** Finding talent is hard work. Finding talent in smaller population groups is even more challenging. The leader must personally engage in searching for and developing talent.

CHAPTER 16   CRISIS MANAGEMENT AND LEADERSHIP

It's one thing to be a leader during "business as usual" times. It's quite another to be a leader in times of crisis and upheaval.

WHAT I LEARNED

1. **Visible and engaged leadership is essential during a crisis.** President Obama's message of responding in fifteen minutes was known by all, and the President repeated this point often. The Governors and Mayors repeated this "response time" expectation as well. Brief, key leader messages can stick and make a difference.

2. **Understand the politics of any situation.** Sometimes it is better to work behind the scenes and let the political leaders be the face of the response effort.

3. **Media.** The media can be very helpful in carrying the message of any organization. They must be engaged with the facts, assumptions, and the strategy regularly and often, so they can keep the public informed.

4. **Teamwork, leadership, and the right equipment makes all the difference.** Each member of the team brings different skills and resources to any successful mission.

5. **Communication.** During a crisis, clear, crisp, consistent, easily understandable, and timely communications are essential.

6. **Education.** In any crisis, it is important to educate self, educate team, and educate stakeholders.

7. **Strengths and weaknesses.** A crisis will shine a light on weaknesses in an organization, a plan, or a leader. A crisis also validates strengths.

8.   Leadership matters more than ever in a crisis. A leader must have both a macro and micro view of the situation. A leader must be adept at situational triage, ensuring the restoration of services in a logical and effective manner that utilizes every available resource.

CHAPTER 17   LEADERSHIP AND PEOPLE—A WINNING COMBINATION

People are the most valuable asset of any organization. Treating people in a way that optimizes their skills by helping them stretch and grow is a prime directive for leadership.

WHAT I LEARNED

1.   Schofield's definition of discipline. Harsh and tyrannical treatment can destroy both an Army and a business if tolerated and allowed to persist.

2.   Talent management. Talent management is important for the growth and success of the individual, the Army, and for a business.

3.   Growing leaders. Giving tactical leaders strategic experiences is essential for their growth and development.

4.   The Warrior Ethos in the Army can apply to business. Never leave a fallen comrade behind in war or business.

5.   Leaders must be visible and present. Leaders must find a way to "be there" during challenging times and make their support and commitment known to their organization.

6. **We are all in the fight. No matter the duties and responsibilities.** Recognize the entire team.

CHAPTER 18   PUBLIC-PRIVATE-PARTNERSHIPS

In today's enterprise climate, success is often achieved through alliances that bring skills and resources to an entire project, thus reducing the risks while improving the benefits.

WHAT I LEARNED

1. **Priority.** If you have a strategic priority in your organization, someone on the team must start each morning knowing that that priority is their primary duty and responsibility.

2. **Experts.** Reach out to the subject matter experts who have experience related to your challenge and learn from their expertise.

3. **Crisis.** At times of crisis, leaders should act quickly, decisively, and make the significant, and at times, revolutionary, changes required. Demonstrate the capacity to recover from a downturn; be resilient.

4. **National programs.** Highly passionate and successful local leaders can implement programs that serve as an exemplar and later scale across the Nation.

CHAPTER 19   TALENT MANAGEMENT

It is not enough to hire the best people. It is even more important to identify and train the future leaders in an organization.

WHAT I LEARNED

1. **Experience at the highest level is important for future senior leaders.** Corporate experience can provide tactical and operational leaders with an understanding of the "big picture," which will help them become more effective leaders at every level.

2. **Mentors play an important role in talent management.** Mentors can make a real and lasting difference in the talent management of young leaders, whether as a trusted adviser, trainer, counselor, or cheerleader.

3. **New experiences broaden the talents of future senior leaders.** Experience matters, but often leaders must work in areas where they have no experience in order to gain the experience that will serve them well later in life.

4. **Do not win a battle at the tactical level only to lose the war at the strategic level.** Failure can happen when junior leaders are too narrowly focused on tactical and operational level objectives at the risk of strategic goals of the overall organization.

CHAPTER 20   YOUR PEOPLE ARE YOUR BRAND'S BEST AMBASSADORS

WHAT I LEARNED

1. **Duty, Honor, Country, and the importance of values.** The values of an organization, military or civilian, underpin its foundation and support its success from generation to generation.

2. **Sometimes leaders must make key decisions at the critical time, which may be out of step with normal processes.** There are times when it is best to make changes the leader knows are right, and then sort out the policy later.

3. **People in any organization are its best ambassadors.** Whether at junior or senior levels, people within an organization can best market and speak about the company.

CHAPTER 21   FROM BOSS TO MENTOR TO FRIEND: THE POWER OF RELATIONAL LEADERSHIP

WHAT I LEARNED

1. **Relational leadership.** Crisis management demands not only leadership that is tactical, strategic, and collaborative but also leadership that is relational. That means leadership that listens and is empathetic, supportive, and responsive.

2. **Checking-in.** Reaching out to people through personal handwritten notes, birthday or anniversary phone calls, and individual and team recognition creates strong bonds in an organization in both good times and crisis.

3. **Focus on people.** Taking the time to focus on people, as well as on the mission, helps build strong winning teams.

AFTERWORD: WINNING AFTER LOSING IS ALL ABOUT RESILIENCE

WHAT I LEARNED

1. **Resilience.** Resilience is the capacity to recover; it is the effort expended to bend but not break despite trauma, tragedy, adversity or crisis. Ultimately resilience is adapting based on the lessons learned, to bounce back stronger than before.

2. **Communications, teamwork, and alignment of goals.** The keys to success are teamwork and mutual support, clear and consistent communications, and alignment of goals at all levels that focus on the priorities critical to achieving success.

3. **America always bounces back.** From the founding of our Nation, we have a great history of bouncing back from difficult times. Whether it was the Spanish Flu, the Great Depression, two World Wars, terrorist attacks, or natural disasters, America has the talent, the resources, the leadership, and the will necessary to bounce back better no matter the challenge, adversity, or crisis.

PHOTOS

From top row to bottom and left to right are: Peter J. Bostick, Fumiko M. Bostick, Aunt Cita Bostick, cousin Linda Bostick, oldest brother Michael C. Bostick, cousin Claudia Bostick , Thomas P. Bostick, brother Anthony C. Bostick, sister Katherine A. Bostick, and cousin Neil J. Bostick. Our family has been resilient through many transitions in life.

After my unsuccessful attempt to obtain a Congressional nomination for West Point, Brigadier General (Retired) George D. Wahl (West Point class of 1917) encouraged me to seek a Presidential Nomination. I was successful in receiving a Presidential Nomination reserved for children of members of the military. Without a nomination to West Point, my plan was to attend the local community college.

Although my father served in combat both in Korea and Vietnam, he never spoke about his military experience. Before I departed for West Point, my parents made this cake. At the time, I did not understand that my father was sending a subtle message that I should become an Infantry Officer, as demonstrated by the cross rifles on the cake.

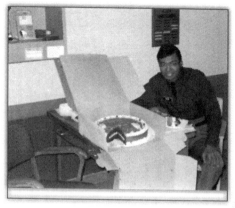

Coming from California, I was a long way from home and I did not have any family travel to see me at West Point until my mother attended graduation. However, my parents always sent a large birthday cake. This gave me the idea of recognizing birthdays for my troops in Germany who were far from home.

I was honored to be a member of the West Point 150-pound Intercollegiate Football Team. Many of my teammates have remained in close contact. Sadly, we lost one of our best, Dexter Curtis "D.C." Adams, #22, in a plane crash in 1997. Teammates created the D.C. Adams Memorial Award, which is presented to the most outstanding second classman on the team. D.C. and I (#80) were always close friends and in the same company, A-2.

1982 U.S. Army Europe Championship Softball Team helped Bravo Company 54th Engineer Battalion win the U.S. Army Best Maintenance Company Award, and they traveled all over Europe developing a foundation of trust.

To my surprise, many soldiers in the company purchased window boxes and decorated their outside windowsills based on my wife Renee's note to the troops.

The Bravo Company Troops enjoyed their two-layer, chocolate fudge birthday cakes—Renee baked 172 birthday cakes while I was in company command.

Bruce Takala (second row, second in from left) and Tom Bostick (first row, second in from the right) were two of three officers on the 1983 All-Army Powerlifting Team. We would both go on to teach at West Point and coach the cadet powerlifting team. Bruce taught nuclear physics and I taught mechanical engineering.

From left to right, Major Paul Christopher, Cadet Wallace Putkowski, Captain Tom Bostick, Cadet Kevin Hartzell, and Cadet Melanie Rowland. West Point won the Collegiate National Powerlifting Championship in 1988.

The 1st Engineer Battalion runs along the Custer Hill loop for physical training April 26, 1996. Afterward, Major General House, the 1st Infantry Division Commander, presented the battalion with the Commander's Cup Trophy based on unit participation and team performances.

The 1st Engineer Battalion leadership team that fought the Idaho Fires in 1994. We also had support from an unexpected friend, Karl "The Mailman" Malone.

The 1st Engineer Battalion Annual Engineer Ball. From left to right, Renee Bostick, Thomas P. Bostick, William B. Gara (first Engineer Commander at Normandy Beach), Lieutenant Colonel and later Brigadier General John R. McMahon (70th Engineer Battalion Commander), and Cathy McMahon.

The White House Fellows Class of 1989–1990 meeting with President George H. W. Bush. From left to right : Michael D. Klausner, Thomas P. Bostick, Daniel B. Poneman, Wayne Tuan, Antonio M. Angotti, Gregory P. Hess, Leigh Warner, Barry R. McBee, Joyce J. Rayzer, John W. Danaher, John McKay, Robert G. Marbut, John W. Orrison.

1st Cavalry Division soldiers land on a new strategic airfield at Tuzla, Bosnia-Herzegovina. Another seemingly impossible mission accomplished in less than ninety days led by the efforts of Air Force Master Sergeant Patrick Daize.

The new Davidson-Style Sea Huts were designed by 2nd Lieutenant Ross Davidson to move troops out of tents. General Larry Ellis envisioned secured facilities with unit integrity, covered walkways, and easy access to showers and latrine facilities.

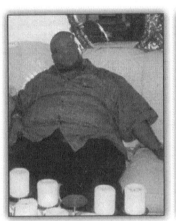 

U.S. Army Private Roderick Evans lost 230 pounds to enlist in the Army. (Photo courtesy Elaine Wilson, Fort Sam Houston, Texas)

Photos

In 2005, Ms. Jackie Purrington served with me in Iraq as a civilian and later became a U.S. Army commissioned officer. She returned to Iraq as a Second Lieutenant Engineer Officer. Courtesy of U.S. Army Recruiting Command, October 2006, Volume 58, Issue 10.

The U.S. Army engaged in many marketing efforts, including the National Hot Rod Association (NHRA) as well as NASCAR, Professional Bull Riding, and Radio and Television to engage America regarding the benefits of serving in the Army. Photo courtesy of U.S. Army Recruiting Command May 2006, Volume 58, Issue 5.

Future Heisman Trophy Winner, Tim Tebow played in the 2006 U.S. Army High School All-American Bowl. Photo courtesy of U.S. Army Recruiting Command February 2006, Volume 58, Issue 2.

My handwritten 3-star notes were often framed by soldiers.

My Executive Officer, Major Cindy Atkins, supported my efforts to teach engineering and leadership to the 4th grade students at Randolph Elementary School where my wife, Renee, was the Principal.

Corporal Simranpreet Lamba, who received a religious accommodation as a Sikh to serve in the U.S. Army carries the guidon for
his platoon in Company A, 3rd Battalion, 34th Infantry Regiment
during Basic Training Graduation.

After speaking with West Point cadets in a discussion about the
importance of their service and diversity in our Army, I met a young
African-American boy and his father. The young boy told me that
he wanted to be a soldier, just like one of these cadets.

May 22, 2012: Hurricane Crisis Planning Guidance where President Obama directed that federal agency leaders and their teams prepare for a hurricane that might take out our electricity, shut down our banking systems, and flood the northeast coast. Hurricane Sandy made landfall on the northeast coast on October 22, 2012.

The U.S. Army Corps leaders met with NYC Mayor Bloomberg early in the response to Hurricane Sandy.

2014 Tuskegee University Commencement Address.

During several opportunities to speak at the United Nations, I was able to engage with key leaders such as UN Secretary-General Ban Ki-Moon and former Prime Minister of South Korea, Han Seung-soo.

Assistant Secretary of the Army for Civil Works, Ms. Jo-Ellen Darcy, and General Bostick visit the troops in Afghanistan.

General Bostick rallies the troops at the steps of the U.S. Capitol building before a bike ride with Wounded Warriors who exemplify the meaning of resilience.

My White House Fellows classmates nominated me for the John Gardner Legacy of Leadership Award. General Colin Powell, also a recipient, presented the award to me on October 25, 2015, at the 50th Anniversary of the White House Fellows Program.

Renee, Joshua, and I celebrate our last Christmas at our quarters at Fort McNair before my retirement ceremony on May 19, 2016 at Ft. Myer, Virginia.

A final formation with the greatest and most resilient team ever—the United States Army.

ABOUT THE AUTHOR

Lieutenant General (Retired) Thomas P. Bostick is an accomplished Senior Executive with more than thirty years of experience in both the public and private sectors. Most recently, he served as the Chief Operating Officer and President of Intrexon Bioengineering (NASDAQ: XON, now Precigen: PGEN). Bostick led Intrexon operations for three health business units and six bioengineering and R&D divisions with over one thousand employees (seven hundred PhD and MS degree scientists) in developing better DNA solutions for energy, health, environment, and food products. He led the restructuring of the company with experience in several merger and acquisition efforts where he worked closely with multiple strategic investors, as well as private equity and venture capital firms, leading to the successful sale of several Intrexon assets. He serves as a Director on the Boards of CSX

(NASDAQ: CSX), Perma-Fix (NASDAQ: PESI), HireVue, American Corporate Partners (ACP), and Streamside Systems.

Bostick served as the 53rd Chief of Engineers and Commanding General of the U.S. Army Corps of Engineers (USACE), where he was responsible for most of the Nation's civil works infrastructure and military construction. He led the world's largest public engineering organization specializing in construction, environment, and water resources management, including seven hundred soldiers and 34,000 civilians with a twenty-five-billion-dollar program in over 110 countries. Under his leadership, USACE was the most improved agency in terms of employee satisfaction in the entire federal government, and the only federal agency with a successful financial audit during his tenure. An expert in crisis response, Bostick controlled the nuclear codes during the tragic events of 9/11, and he led the nation's recovery effort following Hurricane Sandy.

Bostick also served as the Director of Human Resources for the entire U.S. Army with a sixty-two-billion-dollar budget for one million soldiers and 330,000 civilians. He was on the committee that reviewed the law referred to as "Don't Ask, Don't Tell," which was ultimately repealed. He spearheaded the effort to improve opportunities for women and minorities as well as expanding religious accommodations. Bostick was sent from Iraq to command U.S. Army Recruiting Command after it failed its 2005 annual mission. He significantly changed the organization by moving from individual to team recruiting, and as a result, the Army was successful in reaching its recruiting goals during his four-year tenure. He helped create a new Army Slogan, "Army Strong," and led marketing efforts with NASCAR, NHRA, Professional Bull Riding, as well as media such as radio and television.

Bostick was second-in-command of the 1st Cavalry Division and deployed over twenty-five thousand soldiers and their equipment into Iraq. He then commanded the U.S. Army Corps of Engineers Gulf Region Division. He was asked to accelerate a construction program that had only obligated $1.5 billion of an eighteen-billion-dollar program. Within one year, Bostick's team obligated over $11 billion in construction to support Iraq.

Bostick also served as an Associate Professor of Mechanical Engineering at West Point and was a White House Fellow (Class of 89-90), serving as a special assistant to the Secretary of Veterans Affairs. He is a 1978 graduate of the U.S. Military Academy and holds Master of Science Degrees in both Civil Engineering and Mechanical Engineering from Stanford University, and a PhD in Systems Engineering from George Washington University. He was awarded an honorary Doctorate Degree following his 2014 commencement address at Tuskegee University.

He has appeared on CNN, FOX News, CBS Evening News, *The Daily Show*, and has testified before Congress thirty-three times. Bostick has spoken extensively on resilience, including remarks at the United Nations, the Risk Analysis World Congress in Singapore, and the National

Academy of Sciences. Bostick was selected as a member of the National Academy of Engineering, the National Academy of Construction, and the George Washington University School of Engineering and Applied Science Hall of Fame for his many contributions to engineering.

A PERSONAL CLOSING NOTE

"Our lives begin to end the day we become
silent about the things that matter."
~DR. MARTIN LUTHER KING, JR.

The killing of George Floyd has caused the world to look at racism and the impact it is having on the black community. While my own experiences with racism pale in comparison to others, they demonstrate that even in organizations that are laser focused on diversity and inclusion, such as the U.S. Army, challenges remain. Additionally, if leaders at my level have challenges, then those who are not in positions of authority most certainly face racial hatred much more severely.

A SOLDIER'S EXPERIENCE

My son and I were shopping at a military base near Stanford University where my son was an undergraduate student. We decided to shop for some items my son would need at school. I was a two-star General in civilian clothes. We parked our rental car at a spot that had a sign that read, "Colonel and Above." Another car pulled up beside us and parked in the other "Colonel and Above" parking spot. I was working on my phone answering emails when my son said, "Dad, we're in trouble." I replied, "Why?" He said, "See that lady who stepped out of the car next to ours?" "Yes," I replied. Joshua said, "I bet she is calling the police." I asked, "Joshua, why do you think that?" To which Joshua replied, "Two guys that look like us parking in a 'Colonel and Above' space? I'm sure she thinks we're not supposed to be here." I said, "Joshua, there is no way she is calling the police." I stepped out of the car and asked the woman if I could be of assistance. She yelled, "No!" I asked, "Would you like to see my identification card?" She forcefully said, "No, you can show it to the police. You're illegally parked." My son and I waited for the police.

When the military police arrived, I showed them my military identification card. The police were embarrassed and apologetic. I asked, "What exactly did the woman say that led you to believe she was correct in saying that I was illegally parked?" They gave no answer. I knew the answer could only be how we looked. I asked the police for the name of the senior leader at the base. They mentioned the name of a Colonel. I then said, "Okay, I would like to meet with him, your head of police, and the post Equal-Opportunity Officer. I have some shopping to do, and then we can meet in a couple of hours."

The police said, "But Sir, it's Saturday." I replied, "Well, it is Saturday for me, too. I will see you in two hours." During the meeting I told the group that the police should not act on a report about illegal parking when there is no evidence of illegal parking. I was about to close the meeting when Joshua reminded me of another point. I said to the group, "My son noticed that the car of the woman who parked next to us, the same woman who had called the police, has an expired vehicle registration. I would like you to find out who she is, let her know that she is illegally operating her vehicle on base, and that she is banned from entering the base until she corrects this matter."

A FAMILY'S EXPERIENCE

Our family always enjoyed going home to a small city in California for vacation. We never tired of our visits there. We often stayed at a hotel on a nearby military base during our vacations in California. My son would drive from Palo Alto to meet us at the military base. Joshua has very curly, bushy hair, and his car was old and loaded with many of his books and clothes from school. One might think he was living out of his car even though he was earning his undergraduate and graduate degrees from Stanford University. When Joshua approached the gate at the military base where we were staying, he always received strange looks and had to answer a lot of questions. Though he had his military identification card, he was asked questions like, "Whose car is this?" because the car had a one-star decal on the window. Rather than cause any issues for Joshua, who is very sensitive to racism, we started a procedure where Joshua would call his mother or me when he was approaching the gate and one of us would walk outside the gate, move into the driver's seat, and drive the car onto the military base. We went through this procedure each time we visited this military base. We never had an issue with this approach.

Several years later during one of our visits, I was in the gym and Joshua called saying that he was approaching the gate of the military base. I told Joshua I was at the gym and his mother was taking a nap. Joshua said he would just try driving onto the base on his own. A few minutes later, Joshua called me saying that he was at the front gate and needed my help. I ran from the gym to the gate and saw three policemen with their guns drawn and aimed at my son who was still in the driver's seat. One of the policemen was holding my three-star post access card for our car. I asked the policemen what the problem was, and he said that my son had a three-star placard in a car that was obviously not his. I pulled the placard from the policeman's hand and said, "The placard belongs to me, as does the car, and this is my son." Then I asked, "What exactly did my son do wrong?" Dead silence. I asked the policemen to put down their weapons. I said that I wanted to meet with their boss and the senior person from the base. I met with the leadership; they were very apologetic for what had happened and assured me that it would not happen again. At the meeting, I reinforced the need to treat people with dignity and respect and not make assumptions based on the color of a person's skin. We never had an issue again.

LEADERS MUST LEAD

Again, with black people being killed in our streets today, these examples pale in comparison. However, they demonstrate that with racism, one's rank or education matters little. Racism is pervasive and systemic. But it is leaders at every level who act in response to incidents of racism, wherever it occurs, who will be agents for change. Given my seniority in the military, I was able to correct some injustices, not just for me, but for others. But junior minority members of the military and their families, along with those in civilian organizations and other parts of government, have similar issues and are often not able to mitigate the challenges they face.

While it is my experience that most Americans are not racist, racism and racial bias persist. For systemic change to occur, leaders must address the culture that allows racism and racial bias to occur. The type of long-lasting change that is needed requires open dialogue, understanding of the issues, and constant communication. It also depends on diversity, inclusion, and talent management, such that more leaders of all ethnicities and genders are in positions of responsibility at every level in an organization in our military, government, businesses, and education. Again, leadership can make a difference through aggressive and targeted talent management as discussed in this book. I believe that education at an early age, validating and reinforcing the importance of our diversity, is essential.

The tragedy of George Floyd and the movement that has been created by people of all ethnicities across the world offer hope that change will finally occur. Children are not born racist. They learn it. So, we have a chance to ensure that the next generation of children do not learn the racism that plagues our country and many parts of the world. Martin Luther King, Jr. had hoped that his children would "one day live in a nation where they will not be judged by the color of their skin, but by the content of their character." We remain on a long journey to achieve this dream, but the path to success is through our children. Let them continue to demonstrate peacefully for as long as it takes. They will show us the way.

"You may choose to look away, but you can never again say that you did not know."
~WILLIAM WILBERFORCE

INDEX

NUMBERS

1st Armored Division, 267
  Bosnia and Herzegovina, 42,
    44, 310
  Operation Allied Force
    (Albania), 62
  Operation Joint Guardian
    (Kosovo), 62
1st Battalion, 141st Infantry
  Regiment (WWII),
  173–174
1st Brigade, Engineer task force,
  35
1st Cavalry Division, 267
  1-7 Cavalry Squadron, 79–80
  4th Infantry Division rivalry,
    79–80
  activation, 78
  casualties, 84–85
  First Team, 78–87
  Iraq, deployment to, 80
  Port of Rijeka and, 52–56

  Tuzla Airfield, 57–61, 369
1st Engineer Battalion, 18
  1st Infantry Division, 28
  Annual Engineer Ball, 368
  Battle of the Bulge, 36, 37
  Commander's Cup Trophy,
    367
  history, 18–19
  Idaho Fires, 18–21, 368
  Malone, Karl "The Mailman,"
    19, 22–24
  Normandy landings, 36–37
  NTC (National Training
    Center), 30–33
  sports equipment, 21–24
1st Infantry Division, 267
  Big Red One, 319n
  Fort Riley, Kansas, 28–29
4th Infantry Division, 78–79,
  85–87
  1st Cavalry Division rivalry,
    79–80

Fighting Fourth, 79
history, 324n
Iraq, deployment to, 80
Ivy division, 79
Steadfast and Loyal, 79
9/11, resilience lessons, 301
54th Engineer Battalion
(Wildflecken, Germany),
2, 267
Organizational Day, 164
141st Infantry Regiment
(WWII), 173–174

A

A players, 14, 19, 318n, 338
AAR (After-Action-Review), 82,
87, 344
Adams, Dexter Curtis "D.C.,"
363
Afghanistan visit, 376
African-American soldiers, 172
Tuskegee Airmen, 177–179
agile and open, 24
Aide-de-Camp interviews,
208–210
airfield rebuild in Iraq, 105–109
All-Army Power Lifting Team
(1983), 180, 366
alternative options, 13–14,
337–338
Anderson, Joe (Lt. General),
137
Angotti, Antonio M., 369
Army Achievement Medal, 35,
321n
Army Strong motto, 167–169
artillery support
calling for, 31–35

Field Artillery Simulation
Center, 32–35
Granite Mountain, 32–35
asking questions, 63, 342
Assistant Division Command
for Maneuver (1st Cavalry
Division), 81
Assistant Division Command
for Support (1st Cavalry
Division), 81
Atkins, Cindy (Maj.), 372

B

Ballard, Joe (Lt. General), 272
Balocki, James, 233
Ban Ki-Moon, 251–252, 375
Battle of the Bulge, 1st Engineer
Battalion, 36
BDU (battle dress uniform),
281–284
beer ball, 5–6
Behncke, Ted (Lt. Col),
124–126
big picture thinking, 246
birthday celebrations, 160–162,
363, 365
blame game, Softball Strategy
and, 4–5
Bloom, David J., 326n
Bloomberg, Michael, 226–229,
374
BMO (Battalion Maintenance
Office), 2
Bosna, Tuzla Airfield, 56–61
Bosnia, SEA hut, 42–43
Bostick, Anthony, 160, 361
Bostick, Cita, 361
Bostick, Claudia, 361

Bostick, Fumiko M., 361
Bostick, Joshua, 377
Bostick, Katherine A., 160, 361
Bostick, Linda, 361
Bostick, Michael C., 160, 203, 361
Bostick, Neil J., 361
Bostick, Peter J., 160, 361
Bostick, Renee, 377
  1st Engineer Battalion Annual Engineer Ball, 368
  birthday cakes, 160–162, 365
  flower boxes, 163, 364
  Randolph Elementary School, 372
Bostick, Sidney C. (M. Sgt.), 362
  on Vietnam, 279–280
Bostick, Thomas P. (photos)
  1st Engineer Battalion Annual Engineer Ball, 368
  150-pound Intercollegiate Football Team, 363
  1983 All-Army Powerlifting Team, 366
  1988 Collegiate National Powerlifting Championship, 367
  2014 Tuskegee University Commencement Address, 375
  Afghanistan visit to troops, 376
  birthdays for troops in Germany, 363
  Bravo Company Troops birthday cakes, 365
  with Colin Powell, 377
  Custer Hill loop run, 367
  family at Fort McNair, 377
  family on porch, 361
  final formation, 377
  Idaho firefighting team, 368
  with Michael Bloomberg, 374
  Presidential Nomination, 362
  Randolph Elementary School, 372
  UN, with Ban Ki-Moon and Han Seung-soo, 375
  USAREUR Championship Softball Team, 364
  with West Point cadets, 373
  White House Fellows Class of 1989-1990, 369
  Wounded Warriors bike ride, 376
branding, 275–276
  BDU (battle dress uniform), 281–284
  NASCAR and, 281
  NHRA (National Hot Rod Association), 281
  PBR (Professional Bull Riding Association), 281
  "thank you for your service," 283–284
  uniform in public, 280
  U.S. Army All-American Bowl, 281
Brennan, John, 214
Brooklyn-Battery Tunnel, Hurricane Sandy, 221–222
Burcham, Margaret (Brig. Gen.), 332n
Burnett, Leo, 266
Bush, George H.W., 369

C

caring phone calls, 290–291
CASAs (Civilian Aides to the Secretary of the Army), 136–137
certifications, 39
challenges
accepting, 199
crisis-generated, 217–218
change
positive, pursuit, 155, 349
reflection and assessment, 141, 348
resistance to, 130–131
timing, 140, 347–348
Chiarelli, Peter W. (Gen), 323n–324n
Chief of Engineers position, 196, 216
Chisolm, Shirley, 205
Christie, Christopher, 234–235
Christman, Dan (Lt. Col.), 271, 318n
Christopher, Paul (Maj.), 180, 367
Clinton, William (Bill), 43
Collegiate National Powerlifting Championship (1988), 367
combat uniform, 282–283
Commander's Cup Trophy, 367
Commanding General's All-Star Advisory Council, 137–138
competition, 340
goal achievement and, 39
recruiting and, 119–121

Comprehensive Review Group, Don't Ask, Don't Tell, 185–190
congratulatory phone calls, 291–292
Congressional Gold Medal, Japanese-American soldiers, 176–177
contributors, families as, 169, 350
Corbin, Starr, 243
Corvette purchase, 98–99
counterintuitive decision making, 87, 344
COVID-19, resilience lessons, 303–304
Cox, Ken (Maj. Gen), 226, 232
crisis management/leadership, 235, 264, 354–355. See also Hurricane Sandy
communication, 236
crisis communication, 224–226
crisis leadership, 226–229
education, 236
importance, 236
media management, 229–231, 235
national response and local needs, 218–219
Obama meeting, 213–214
politics of the situation, 235
relational leadership, 289
strengths and weaknesses, 236
talent management and, 231–233

cultural awareness, 114,
    346–347
Cuomo, Andrew, 230–231
Custer, George Armstrong, 80

D

Daize, Patrick (M. Sgt.), 59–60,
    256, 322n, 369
Dalrymple, Jack, 262
Danaher, John W., 369
Darcy, Jo-Ellen, 376
DAS (Directory of Army Staff),
    290–291
data gathering, 199
Davidson, Ross (2nd Lt.), 370
    Medical Service Corps,
        47–49
    SEA huts, 45–47
Davidson-Style SEA Hut,
    41–42, 44–47, 370
D.C. Adams Memorial Award,
    363
decision making, counterintui-
    tive, 87, 344
deep vein thrombosis, 90–94,
    326n
    deployment after, 95–97
DeFour, Wilfred, 177
deliberate defense, 36, 320n
DeLuca, Peter "Duke" (Brig.
    Gen.), 233
Derwinski, Edward J., 266, 271
discipline, Gen. John M.
    Schofield, 239–240, 249, 355
Distinguished Service Cross, 1st
    Cavalry Division recipients,
    78

diversity, 171
    African-American soldiers,
        172
    integration of military, 178
    Japanese-American Nisei, 176
    Japanese-American soldiers,
        172, 173–177
    Native American soldiers, 172
    Sikhs in the Army, 194–197,
        373
    strength in, 199, 352
    Tuskegee Airmen, 177–179
    women in the Army, 172,
        179–180, 193–194
Division Warfighter Exercise,
    62–63
Don't Ask, Don't Tell, 184–198
    misquote in press, investiga-
        tion, 189–193
    Navy SEALs, 187–188
    performance and, 188
    Stuttgart, Germany group,
        188–193
downtime, 14–15, 339
dreamer leaders, mission
    impossible and, 42
Durham-Aguilera, Karen,
    214
"Duty, Honor, Country"
    speech (MacArthur), 238,
    277–278, 358

E

electricity restoration in Iraq,
    109–111
Ellis, Larry (Maj. Gen.), 42,
    321n–322n

Division Warfighter Exercise, 62–63

Port of Rijeka, 53–56

SEA huts, 44–47

Tuzla Airfield, 56–61

enlistment age, 148, 152

Erickson, McCann, 266

ESF (Emergency Support Functions), 215

Hurricane Sandy, 215–216

European tours

Lloret de Mar, Spain, 7–8

Nuremberg, 6–7

Rome, 9–12

softball championship and, 9–10

Evans, Roderick (Pvt.), 145–149, 370

Executive Officer positions, 267

experts

success and, 256

trusting, 263, 343

using well, 64, 343

F

failure, challenge of, 2–3

families

Army Strong motto, 168–169

author siblings, 160

birthday cake at West Point, 160

as contributors, 169, 350

individual input, 158

morale and, 163–164

pancakes for dinner, 164–166

recruiting and, 165–166

of team members, 88, 345

visits to West Point, 159–160

Fargo-Moorhead flood diversion project, 261–263

FEMA, Hurricane Sandy and, 217–218, 224–226

Field Artillery Simulation Center, 32

Fighting Fourth (4th Infantry Division), 79

fire fighting, 18–21, 368

sports equipment, 21–22

First Break All the Rules, What the World's Greatest Managers Do Differently (Buckingham), 125–126

First Team (1st Cavalry Division), 78–87

flower boxes for windows, 163, 364

focus, success and, 256–257, 345

Fort Hood, casualties, 1st Cavalry Division, 84–85

Fort Irwin, NTC (National Training Center), 18

Fort Leonard Wood, 233

Fort Ord, 257

Fort Riley, wrestling, 28–29

friendship, 85–87

Fugate, Craig, 214, 217

future leaders

characteristics, 201–211

mentors, 205–207

passion and, 203–204, 353

persistence and, 204–205, 353

role models, 203, 353

talents, 208–211

G

Gara, William B. (Lt. Col),
37–38, 368

Gates, Robert, 186, 191

geographically dispersed workforce, 104

global teams, 104–105

GOMO (General Officer
Management Office),
118–120

*Good to Great: Why Some
Companies Make the Leap and
Others Don't* (Collins), 14

Granite Mountain, artillery
simulation and, 32–35

Grassroots Community
Advisory Board, 135–137

Green Zone (Baghdad),
247–248

Gridley, Richard (Col.), 216

growth of leaders, 250

H

Hadfield, Chris, 242

Ham, Carter (Gen.), 186, 188–
193, 331n

Han Seung-soo, 375

Hartzell, Kevin (Cadet), 367

hasty defense, 33–34, 320n

Heitkamp, Heidi, 262

heroes, 155, 349
recognition, 181, 351

Hertling, Mark (Lt. Gen.), 331n

Hess, Gregory P., 369

Hicks, Paul (Capt.), 243–244

higher loyalty construct, 87

history, using well, 39, 340

Hitler, Adolph, Lost Battalion
(WWII), 174

Hoeven, John, 262

Holley, Danny, 257–259

Honore, Russel L. (Gen.), 228

House, Randolph W. (Maj.
Gen.), 319n

HR (human resources)
centralization, 138–139
Don't Ask, Don't Tell,
184–186

Hurricane Crisis Planning
Guidance meeting, 213–
214, 374

Hurricane Katrina, 215

Hurricane Sandy, 214–215
Bloomberg, Michael, 226–
229, 374
Brooklyn-Battery Tunnel,
221–222
crisis communication,
224–226
crisis leadership, 226–229
ESF (Emergency Support
Functions), 215–216
FEMA and, 217–218
media management, 229–231
national response and local
needs, 218–219
New Jersey blocked, 219–220
power grid issues, 223–224
resilience lessons, 301–302
talent management and,
231–233

Hutchinson, Deborah, 94

Hutchinson, Michael, 94

Hutchinson, Ray Joseph (Spec.),
94–95

I

Idaho Fires, 18–21
inclusiveness, team-building, 114
individual talents, 14, 338
inexperienced team members,
    50, 341
instincts, following, 115, 347
Institute for Water Resources,
    255–256
integration of military, 178
intentionality, 141, 348
international teamwork,
    103–104
  gravel challenge, 105–109
  local culture, 108
  power grid challenge,
    109–111
  share the success, 111–114
Iraq
  1st Cavalry Division, 80
  4th Infantry Division, 80
  airfield rebuild challenge,
    105–109
  electricity restoration,
    109–111
  reconstruction, 104
  school rebuilds, 112–114
Ivy division (4th Infantry
    Division), 79
Iwata, David, 137

J

Japanese-American Nisei, 176
Japanese-American soldiers, 172
  442nd Regimental Combat
    Team, 173–174
  Congressional Gold Medal,
    176–177
Lost Battalion (WWII),
    173–174
John Gardner Legacy of
    Leadership Award, 377
Johnson, Dan (Ensign), 68–74
Johnson, Jeh, 186, 331n
Johnson, Sallie, 72
Johnson, Wallace, 72
Johnson, William, 177
Jones, James (Capt.), 128, 325n

K

Klausner, Michael D., 369
Kosovo, 61–63
  Operation Joint Guardian,
    62–63
Kuwait, 88

L

Lamba, Simranpreet (Spec.),
    331, 373
Lasorda, Tommy, 137
leaders. *See also* future leaders
  growth, 250, 355
  mentors, 205–207
  persistence, 204–205, 353
  resilience, 299–300
  support for, 198, 351–352
  talents, 208–211
  with vision, 49–50, 340
  visionary, 63, 342
leadership
  experience levels, 75, 357
  modesty and, 75, 344
  people-centered, 65–75
  performance and, 183–198
  qualities in oneself, 202–203
  relational, 287–296, 358–359

servant leadership, 66
supportive, 198, 351–352
Lew, Jack, 214
listening, 199, 348, 352
Lloret de Mar tours, 7–8
local culture, 114
teamwork and, 108, 114, 346–347
Long Gray Line, women in, 179–180
Lost Battalion (WWII), 173–174

M

MacArthur, 276–280
MacArthur, Douglas (Gen.), "Duty, Honor, Country," 238, 277–278
MAIT (Maintenance and Assistance Inspection Team), 2, 318n
Saturday Certification Program, 2
Malone, Karl "The Mailman," 17
1st Engineer Battalion (Boise City, Idaho), 19–23
U.S. Army High School All-American Bowl Football Game, 23–24
Mann, Lisa Reyn (Maj.), 332n
Marbut, Robert G., 369
McBee, Barry R., 369
McHugh, John, 191
McKay, John, 369
McMahon, Cathy, 368
McMahon, John R. (Brig. Gen.), 368

Medal of Honor, 1st Cavalry Division recipients, 78
media, in crisis, 229–231
Medical Recruiting Brigade, 128–131, 328n
mentors
talent management and, 273, 357
unexpected, 205–207, 353
Meurer, Fred, 256–257
Ford Ord, 258
Monterey Model, 259–260
Military Housing Privatization Act, 259
military strategy vs business strategy, 237–249
minefields, 36–38
Mission First, People Always, 66, 238–240
mission impossible, 42
Ellis, Larry and, 43–44
modesty, leadership and, 75, 344
Monterey Model, 259–260
morale, family and, 163–164
MOS (military occupational specialties), 125–126
Mullen, Mike (Admiral), 292
Myers, Frank, 243, 245

N

NACCS (North Atlantic Coast Comprehensive Study), 301–302
Nagaoka, Minoru, 174–175
Nagl, John, 238
Napolitano, Janet, 214
NASCAR, 281, 371
national programs, 264

Native American soldiers, 172

Navy SEALs, Don't Ask, Don't Tell and, 187–188

NCR (National Capital Region), 266

NHRA (National Hot Rod Association), 281, 371

Nishizawa, Eric, 137

NMCC (National Military Command Center), 267

Normandy, 1st Engineer Battalion, 36

North Texas Chapter of the League of United Latin American Citizens, 137

NRF (National Response Framework), 215–216

NTC (National Training Center), 28
  Fort Irwin, 18
  OPFOR (Opposing Force), 30

Nuremberg, Wall, 6–7

O

Obama, Barack
  Don't Ask, Don't Tell repeal, 193
  Hurricane Crisis Planning Guidance meeting, 213–214, 374
  Hurricane Sandy, 218–219

obesity, 146–147

Odierno, Raymond T. (Gen.), 324n

OEMA (Office Economic and Manpower Analysis), 139

Operation Allied Force (Albania), 1st Armored Division, 62

Operation Joint Guardian (Kosovo), 1st Armored Division, 62–63

OPFOR (Opposing Force), 30, 320n

Option 3, 10–12

Orrison, John W., 369

Owen, Paul (Col.), 226

P

P3 (pubic-private partnership), flood risk management, 254–263

pancakes for dinner, 164–166

Panetta, Leon, 194

passion
  future leaders and, 203–204, 353
  of individuals, 50, 342

PBR (Professional Bull Riding Association), 281

Peck, Gregory, as MacArthur, 276–280

Pelley, Scott, 230

people-centered leadership, 65–75

perception in a crisis, 226–229

performance, 199, 352
  Don't Ask, Don't Tell and, 188
  leadership and, 183–198
  talent assessment and, 181

Perot, Ross, 136

persistence, future leaders and, 204–205, 353

personal resilience, 100, 300

Peterson, Collin, 262

Petraeus, David, H. (Maj. Gen.), 325n

Poneman, Daniel B., 214, 369

Port of Rijeka, 52–56, 322n

positive change, pursuit, 155, 349

post-mortems with team, 82

Powell, Colin (Gen.), 377

powerlifting team, 366, 367
  coach, 180
  Collegiate National Powerlifting Championship (1988), 367

Prescott, Tony (Sgt.), 325n

Primer Power Battalion, Sikhs in the Army, 196–197

priorities, 263

project funding, 251
  speed of completion, 252–253

public-private partnership, 251–264, 356

pulmonary embolism, 90–94, 326n
  deployment after, 95–97
  General's story, 97–98

Purrington, Jackie (2nd Lt.), 149–152, 371

Putkowski, Wallace (Cadet), 367

Q

questions, asking right ones, 63, 342

R

Rainey, Mike, 243

raising the bar, 63

Rayzer, Joyce J., 369

RCI (Residential Communities Initiative), 259–260, 334n

recognition of team, 248–249, 351

recruiting, 117. *See also* USAREC (U.S. Army Recruiting Command)
  competition and, 119–121
  as family affair, 165–166
  GOMO and, 118–120
  team recruiting, 123–126
  USAREC, 119–124

reflection, 141, 348

relational leadership, 287–289, 296, 358–359
  caring phone calls, 290–291
  checking-in, 296
  congratulatory phone calls, 291–292
  crisis and, 289
  focus on people, 296
  personal attention, 292–293
  personal celebrations, 293–294
  personal notes, 294–295

resilience, 297–298, 359
  9/11 and, 301
  COVID-19 global pandemic, 303–304
  Hurricane Sandy, 301–302
  leaders, 299–300
  leadership and, 75, 343
  personal, 100, 300
  unreachable goals, 302
  Vietnam War aftermath, 302–303

responsiveness, 249

Ricks, Tom, 230

role models, 75, 343, 353

future leaders and, 203, 353
Rowland, Melanie (Cadet), 367

S

Saturday Certification Program, maintenance, 2
Savre, Kent (Brig. Gen.), 232
Schofield, John M. (Lt. Gen.), 239–240, 249, 355
Schwarzkopf, Norman (Gen.), 167
Scott, Bruce (Maj.), 271
SEA (Southeast Asia) huts, 44, 370
  Davidson, Ross, 45–47
selfless service, 100, 345
Semonite, Todd (Lt. Gen.), 233
servant leadership, 66
Sessions, Jeff, 191
setbacks, recovery, 101
Shima, Terry, 174–175, 176
Shineski, Eric K., 267
Sikhs in the Army, 194–197, 373
Silver Star, Vaccaro, Angelo (Cpl.), 154
Sinek, Simon, 317n
Softball Strategy, 3–5
  beer ball, 4–5
  Turnaround Softball Strategy, 6–7
  USAREUR (U.S. Army Europe) Championship, 8
  Wildflecken Softball Championship, 8
Special Assistant to the Secretary of Veterans Affairs, 266

sports, 340
  Commander's Cup Trophy, 29
  equipment, Malone, Karl "The Mailman," 17, 19–24
  importance, 25
  team building and, 38, 340
  as team-building tool, 28–29
  wrestling at Fort Riley, 28–29
star notes, 294–295, 372
strategy, military versus business, 237–249
success, sharing, 111–114
support
  accepting, 100, 346
  for leaders, 198, 351–352
  supportive leadership, 198, 351–352

T

tactical leaders versus non-tactical leaders, 268–271, 357
Takala, Bruce, 366
talent assessment
  future leaders, 208–211
  performance, 181
talent management, 83–84, 242, 250, 265–273, 355, 357
  mentors and, 273, 357
  redirecting, 271–272
Task Force Bostick, 33–35
team members
  families of, 88, 345
  inclusiveness, 114
  player types, 347
  quiet persons, 169, 350
team building

creativity in, 140–141
inclusiveness, 114
A players, 14, 19, 318n, 338
SEA huts, 44–47
Softball Strategy, 3–5
sports and, 38, 340
teams
  balance between individual
    and group, 77
  one voice, 141, 348
  recognition, 248–249
Tebow, Tim, 371
Ten-Run Rule, 9, 11
Tortolano, Astro, 174–175
Toy, Mark (Brig Gen), 232
TRADOC (Training and
    Doctrine Command),
    Sikhs in the Army, 195
training
  certification and, 39
  universal applications,
    24–25
Tuan, Wayne, 369
Tuskegee Airmen, 177–179
  DeFour Wilfred, 177
  Johnson, William, 177
Tuskegee University commence-
    ment address, 179, 375
Tuzla Airfield (Bosnia), 56–61,
    322n
  Daize, Patrick, 59–60

U

unconventional tools, 25
United Nations speaking
    engagements, 375
U.S. Army Accessions
    Command, 167–168

U.S. Army All-American Bowl,
    281
U.S. Army Corps of Engineers
  Command General position,
    196
  flood risk management,
    254–256
  Hurricane Sandy and,
    215–216
  Institute for Water Resources,
    255–256
  Panama Canal, 255
  projects, 216
U.S. Army High School All-
    American Bowl Football
    Game
  Malone, Karl, 23–24
  Tebow, Tim, 371
U.S. Forest Service, 18, 19–20
USAREC (U.S. Army
    Recruiting Command),
    119–120, 327n–328n
  3rd Recruiting Brigade, 124,
    126–127
  change to model, 122–123
  communication changes,
    133–134
  Grassroots Community
    Advisory Board, 135–137
  information technology and,
    140
  Medical Recruiting Brigade,
    128–131
  Milwaukee Recruiting
    Battalion, 124–125
  pocket talking points,
    133–134
  recording cameras, 133

team recruiting, 123–126
USAREUR (U.S. Army Europe)
  formation, 317n–318n
  Softball Championship, 8,
    364

V

V Corps, 318n
  Distinguished Small Unit
    Award, 164
  MAIT team, 2
Vaccaro, Angelo (Cpl.), 152–154
Vietnam War, resilience lessons,
  302–303
VII Corps, 318n
vision
  inexperienced team members,
    50, 341
  leaders with, 49–50, 340
visionary leadership, 63, 342
VTC (Video Teleconference
    Conference)
  Hurricane Sandy, 218–219
  Port of Rijeka and, 54

W

Wahl, George D. (Gen.), 205–
  206, 332n, 362
Ward, Kip (Gen.), 267
Warfighter Exercise, 62–63,
  92–93, 326n
Warner, Leigh, 369
Warrior Ethos, 173–174, 250,
  355
Warrior Transition Brigade, 154
Wehr, Mike (Maj. Gen), 232
West Point

1st Engineer Battalion and,
  18–19
150-pound Intercollegiate
  Football Team, 363
family visits, 159–160
powerlifting team, 180, 366,
  367
Presidential nomination, 206
women at, 179–180
young African-American boy,
  207, 373
White House Fellows Program,
  266, 271–272, 335n, 369
Wildflecken, 5
  birthday celebrations,
    160–162
  Softball Championship, 8
Williams, Arthur (Lt. Gen.),
  267
window flower boxes, 163, 364
Witten, Jason, 136–137
women in the Army, 172,
  193–194
  West Point, 179–180
Women's Armed Services
    Integration Act, 172–173
Women's Army Corps, 172
World War II, Lost Battalion,
  173–174
Wounded Warriors bike ride,
  376
wrestling at Fort Riley, 28–29
Wright, Steven, 70, 72

X–Y–Z

Yamaguchi, Kristi, 29
Zaidi, Ali, 262

Made in the USA
Coppell, TX
16 March 2021

51826874R00238